Electric Life

Urban and Industrial Environments

Series editors: Robert Gottlieb, Henry R. Luce Professor of Urban and Environmental Policy, Occidental College

Bhavna Shamasunder, Associate Professor of Urban and Environmental Policy, Occidental College

A full list of titles in this series is included at the end of this book.

Electric Life

Utility Regulation and the Fight for Energy Democracy

Nikki Luke

The MIT Press
Cambridge, Massachusetts
London, England

The MIT Press
Massachusetts Institute of Technology
77 Massachusetts Avenue
Cambridge, MA 02139
mitpress.mit.edu

The MIT Press would like to thank the anonymous peer reviewers who provided comments on drafts of this book. The generous work of academic experts is essential for establishing the authority and quality of our publications. We acknowledge with gratitude the contributions of these otherwise uncredited readers.

This book was set in Stone Serif and Stone Sans by Westchester Publishing Services. Index by Twin Oaks Indexing. Printed and bound in the United States of America.

Library of Congress Cataloging-in-Publication Data is available.

ISBN: 978-0-262-05197-2

EU Authorised Representative: Easy Access System Europe, Mustamäe tee 50, 10621 Tallinn, Estonia | Email: gpsr.requests@easproject.com

Contents

Acknowledgments vii

Introduction: The Value of Electric Life 1

1 Lighting the Atlanta Way 31

2 Paying for Electric Life 59

3 The Slow Violence of Fossil-Fueled Progress 87

4 Organizing Electricity on and off the Job 113

5 The Work of Finding Work 139

Conclusion: Stealing Back Time to Build the Future 165

Notes 177

Index 245

Acknowledgments

The ideas in this book developed over more than a decade, and I have been lucky to think with and learn from brilliant and kind people along the way. There are many people to whom I owe gratitude for demonstrating the power in and possibility of building collective infrastructures. In Atlanta, many people shared their thoughts and time with me in interviews. Neil Sardana, Wan Smith, Gayla Tillman, Jillian Eller, and Ben Clark welcomed me to the We the Plug Tho squad. The Georgia Stand-Up Policy Institute for Civic Leadership grounded me in the political economy and social movement history of Atlanta. Neill Herring shared his personal files on environmental movements in Georgia. Librarians at the Southern Labor Archives, the Atlanta History Center, and the Georgia State Archives helped me access collections before and during COVID. Melanie Barron, Bri Knisley, June Rostan, and Vonda McDaniel have become important collaborators and friends. Thanks to everyone in the Tennessee Valley Energy Democracy Movement for the model of popular organizing for public power. Becky and Keith encouraged my scholarship amid many other things we had to do and helped me keep this project moving forward. Some of the people who informed my thinking in this book are no longer here. My grandmother, Edith Heidbreder, taught me about the intergenerational work of care. Fran Ansley's generosity and deep commitment to people remains a guiding inspiration for my academic work. I cherish memories of talking about prefigurative politics with Muhiyidin Moye.

I completed my dissertation, which eventually became this book, at the University of Georgia. In Athens, I learned from Hilda Kurtz, Steve Holloway, Susan Thomas, JoBeth Allen, Betina Kaplan, Melissa Perez-Rhym, and U-Lead Athens. Class with Josh Barkan changed the way that I ask questions and prompted me to dig into the historical and legal dimensions of utilities. Matt Huber's enthusiasm for this project and the wonky details of electricity politics helped me to articulate why they matter. Nik Heynen was the most unbelievably patient and generous advisor. He rooted me in an intellectual community in geography and the South. Throughout many moves, Mark Collins and Maria Kaika have remained beloved mentors and friends.

At the University of Tennessee, I am grateful for the camaraderie of LaToya Eaves, Solange Muñoz, Nicholas Nagle, Pam Sharpe, Derek Alderman, Wendy Bach, Kelsey Ellis, and Sally Horn. This work benefitted from the financial support of the National Science Foundation.

Many other geographers and a wider research network have provided constructive criticism, reading recommendations, and inspiration over the years, especially Caroline Keegan, Dani Aiello, Sara Thomas Black, LaToya Eaves, Carrie Freshour, Brian Williams, Aaron Mallory, Theo Hilton, Conor Harrison, Shelley Welton, Elena Louder, Myles Lennon, Kristin Phillips, Sasha Tycko, Ben Postone, Alex Moulton, Rebecca Lave, Patrick Bigger, Johanna Bozuwa, Mijin Cha, Kelly Kay, Marion Werner, Leila Harris, Mukul Kumar, Jonathan McCombs, Richard Milligan, Dean Hardy, Alice Hilton, Imani Scott-Blackwell, Rachelle Berry, Nina Ebner, Akira Drake-Rodriguez, Deepti Chatti, Danya Al-Saleh, Xi Wang, Katherine Hankins, and Will Bryan.

The ideas here were much improved through collaborations. Writing with Matt Huber, Rachel Goffe, and Wan Smith advanced my thinking about electricity, social reproduction, and energy democracy. These coauthored pieces can be found in *Environment and Planning E*, *Environment and Planning A*, and the *International Journal of Urban and Regional Research*. Parts of the introduction, chapter 2, and chapter 4 have been previously published in *Environment and Planning E* and the *Annals of the American Association of Geographers*.

Feedback from many audiences helped me tell a better story. Thank you to Nina Ebner, Mikael Omstedt, Charles Sims, Budhu Badhuri, Alissa Kendall, John Stehlin, Shelley Welton, Ankit Kumar, and Casey Williams for organizing talks and sessions where I received feedback on chapters. Jon Shefner, Lois Presser, and Josh Barkan reviewed early drafts. Nik Heynen read (actually) countless versions. Gabi Lichtenstein, Caroline Hamilton, and Gabe Schwartzman helped with final revisions. At the MIT Press, Beth Clevenger encouraged this project over many years. Anthony Zannino and many others guided me through the production process. The book is much improved from this careful and generous feedback as well as that of three anonymous reviewers. All errors remain mine alone.

In Knoxville (and beyond), the fierce love, deep laughter, and determination of a wider network of friends and family keeps me going. I am grateful every day for Gabe, who is my partner in all things and gave up too many weekends for me to finish writing. To my family, I owe so much, especially my brother Kit, my in-laws, Pat and Steve, and my parents, Kay and Tim. My parents have always encouraged me to be curious. This book is for them.

Introduction: The Value of Electric Life

On a hot and sticky September afternoon in downtown Atlanta in 2019, Georgia Power customers gathered on the sidewalk across the street from the Georgia Public Service Commission. Crowded in the shade of oak trees, the demonstration was organized to protest a proposed monthly fee and rate increase by the regulated electric utility, Georgia Power, that would raise average electricity bills by $16 a month for residential customers. Young people, elders, and faith leaders testified to their everyday needs for electricity to complete homework, cook meals, power life-sustaining medical equipment, and illuminate churches and community spaces. They spoke about the significant burden of high electricity bills and how the hike would affect daily life for Atlantans already struggling to afford the increasing cost of living. Atlanta is among the most unequal cities in the US and has among the highest median energy burdens, which measures the ratio of energy costs relative to income. For residents of the city's wealthier, northern neighborhoods, who might have their utility bills automatically deducted from their bank accounts, the light bill is inconsequential in a month's balance sheet. In historically Black neighborhoods in southwest Atlanta, and across low-income, racially diverse communities in the suburbs, energy bills consume upward of 10 percent of household budgets. This outpaces the 6 percent threshold often used to define affordable energy and more than doubles the metro area median for energy bills that consume 5 percent of household income.[1]

One woman from the Malcolm X Grassroots Movement put it simply: "We ask for $15 an hour and we're told it's not feasible. Now we know this increase [in electricity bills] is coming and there's been no increase on our side."[2] Her statement—in reference to the long history of organizing by low-wage workers for a $15-an-hour minimum wage—condemned the devaluation of labor that leaves working people unable to afford necessities of everyday life like electricity. The electric bill becomes another obligation that forces working-class people to work a second job, take out payday loans, piece together assistance from overextended charitable and public coffers, or borrow from relatives to try to keep the lights on. The demonstrators pleaded with the elected Public Service Commissioners who regulate the investor-owned, monopoly electric utility to deny Georgia Power's request for a rate increase and require renewable energy and energy efficiency programs that address climate change and provide for community needs. At issue was more than the rate increase at Georgia Power. The protest launched a broader critique about how the utility and the public officials who regulate it should operate to serve the public interest.

In the years leading up to and since the protest, public awareness about and participation in electricity regulation has grown in Georgia and nationally. Organizing endeavors to cut through wonky and opaque regulation, legislation, and corporate process that govern electricity to hold utilities accountable to the concerns of everyday people. In this book, I study efforts to change the utility within the broad framework of energy democracy. Energy democracy strives to make governance processes open and responsive to the public to build more equitable and environmentally conscious utilities equipped to fight climate change, provide remedy and reparation for enduring environmental and energy injustice, and ensure a just and equitable energy transition from fossil fuels to clean energy sources for utility workers and communities that live with energy infrastructures.[3] This book contributes to this work by asking how electric utilities came to exert a profound yet hidden influence over our economic and social life, what this control means for our collective futures, and how ordinary people can participate in reimagining how electricity is generated, owned, and operated.

Even as attention on utilities and public service commissions has grown, inertia creates barriers to change.[4] Nearly 70 percent of households in the US get their power from a large, investor-owned utility.[5] The organization of these firms dates to the early twentieth century when utilities were granted monopolies over geographic service territories. In the decades since, utilities and the public service commissions that regulate them have worked collaboratively, forming what Richard Hirsh terms a "utility consensus" organized around economic growth.[6] The particularities of the utility business model dissuaded experimentation, and regulators have largely failed to respond to

Figure 0.1
Demonstrators at a rally to Fight the Hike outside the Georgia Public Service Commission in Atlanta, 2019. Photo Credit: John McCosh.

calls for climate action and environmental and energy justice. Some argue utility regulation is a textbook case of industry capture as regulators defer to utility expertise.[7] Public service commissions are further constrained in their role as economic regulators and have limited authority (or appetite) to confront broader social, environmental, and climate issues.[8]

These limitations have real consequences, not least because utilities are key to mitigating climate change. In 2023, the electric power sector produced 34 percent of the energy used in the US and was the source of about 30 percent of CO_2 emissions.[9] The relative importance of electricity in our energy mix is only expected to increase with the push to electrify everything to minimize greenhouse gas emissions from transportation, cooking, and industrial processes like steelmaking that rely on fossil fuel combustion.[10] Paradoxically for intransigent utilities that have long denied or downplayed their own contributions to climate change, the climate crisis presents a business opportunity. After experiencing flatlining demand since 2008, the US saw its total electricity demand rise in 2022, with data centers, new factories, and electrified transportation coming online. Georgia Power estimates new load growth, about 80 percent of which is driven by data centers, is the largest and fastest growth in electricity consumption since the introduction of air conditioning and could increase Georgia's peak demand nearly 60 percent by

2035.[11] Controversially, the utility has proposed to meet this new demand by building new gas generators while also maintaining coal-fired power plants.

Federal policies such as the 2021 Bipartisan Infrastructure Law (BIL) and the 2022 Inflation Reduction Act (IRA) incentivized alternatives, including a range of clean energy technologies and renewable energy sources such as wind and solar.[12] Renewable energies supplied more than 21 percent of the electric grid nationally and 6 percent in Georgia in 2023.[13] Yet, utilities have been slow to move away from fossil fuels entirely. Decarbonizing electricity is not a simple matter of one-to-one substitution of fossil fuels for renewables. Intermittency and geographic variability of renewables requires storage to maintain twenty-four seven reliability throughout the year. Renewable energies can be integrated into the grid in a range of scales from utility-scale to distributed energies. This flexibility has motivated efforts to democratize access to renewable energy sources by shifting away from the twentieth-century hub-and-spoke model organized around centralized power plants to a more distributed system in which "prosumers" own battery storage, solar panels, smart appliances, and electric vehicles that they use to generate power and manage demand. Changing the grid, which is as much a legal and financial architecture as it is an engineering feat, disrupts incumbent technologies and also upsets the stakeholders wedded to them. The energy transition is a technical question about how to decarbonize the grid *and* a political and social one about how best to provide for our quotidian dependencies on power in electric life.

Electricity is both a vital and banal resource that intertwines in countless, ordinary moments and illuminates our most intimate and mundane daily tasks. Despite its ubiquity, electricity is a peculiar commodity that has limited use in and of itself. Rather, we want electricity for what it can do. Electricity powers classroom computers, hospital X-ray machines, church speaker systems, ATMs, corner store cold cases, streetlights, televisions, and countless other elements of the city. It makes possible economic activity—the production of goods and services—and our social lives.[14] You may be using electricity right now to read under a lamp or on a computer screen, stay warm or cool inside, or to make a cup of coffee. I study the ubiquitous use of electricity in everyday life as *electric life*. I use this term to refer to the functions, spaces, and times that electricity shapes including both economic activity and our social lives beyond the worksite. My analysis of electric life builds on feminist political economists' study of social reproduction, which encompasses the waged and unwaged work carried out by households, communities, and the state to reproduce social and economic relations every day and over generations.[15] This framework attempts to move beyond dominant narratives of "productive" economic activity by highlighting the diverse kinds of work that contributes to making the economy and social life possible, including preparing meals, caring for children, and training future generations of workers.[16] Electric life is a wide lens that holds in tension the diverse influence of electricity on workplaces, homes, government, schools, and the environment and, in moving across them, asks how electricity draws on and transforms social reproduction.

Our everyday dependence on electricity contributes to the political and economic might of utilities, which have become powerful political institutions. Every time we flick on a switch, we consent to the practices of our electric provider. And, in the territories of private utilities like Georgia Power, we authorize our utility to earn a profit from the sale of electricity. In other words, in Georgia, electric life is made possible by and animates electricity capital. Electricity capital is the nexus of state, regulatory, and financial relationships that enables profit from electricity provision.[17] Electricity capital is not purely a creature of private enterprise but results from the intertwined business and state regulatory practices that organize, govern, and make possible accumulation through electricity provision. As much as the utility creates the conditions of electric life, it benefits from regulation that, in the case of Georgia Power, protects its monopoly territory and enables the utility to capture social reproductive labor that individuals and organizations take on to survive within is system.

This collaboration between the utility and the state drives racialized uneven development.[18] To see electricity capital as an infrastructure that powers racialized uneven development draws into focus the contradictions of electricity-powered progress, such as why energy poverty persists alongside shareholder wealth? Why disinvestment continues in the face of urban growth? Why production of bright, clean current generates coal ash impoundments and asthma-inducing pollution? Why do we continue to burn coal and gas that emit climate-warming pollutants even as we see the disastrous effects of extreme weather? Fundamentally, why are these risks that threaten premature death for some an acceptable consequence of progress and how are these otherwise impossible choices made possible by social prejudices that designate the lives of people of color, women, and poor, disabled, and rural people less valuable?

To understand these inequalities requires studying investor-owned utilities within the wider political economy in which they are constituted. I begin from the historical position that our contemporary economic system of capitalism has always been racial capitalism that relies on and reinforces social difference and reproduces inequality defined through the interdependent categories of gender, race, class, nationality, and ability.[19] Understanding inequalities reproduced through electricity capital requires investigating the historical evolution of power between social groups that define how time, labor, and property are distributed, valued, and devalued within Atlanta. It also requires nuanced attention to people's lived experiences and uses of electricity to understand how ordinary customers and workers navigate and contest these trade-offs.

My contention in this book is that the power of the regulated, investor-owned electric utility comes from the everyday lives of its customers. In arguing this, I make three key interventions that expand critical studies of energy and energy justice by centering the people who comprise the utility and who utilities are built to serve. First, I assert that we must consider how cooperation between the state and utilities empowered private companies to build the networks of electric life. Second, I argue that control over the time of electric life is what creates value for the utility. Third, in reviewing how the electric grid is organized and maintained, I ask why the public has empowered a private utility to provide an essential service rather than guarantee its provision as part of a social wage. The history of how this system was built in Georgia can help us collectively envision new ways to organize utilities and the institutions that regulate them to address the intertwined crises of climate change and racial and economic inequality.

Throughout the nearly 150-year history of private electric utilities, organic social movements have emerged from the homes where electricity is used, the worksites where electricity is produced, and the communities affected by pollution from electricity production. These campaigns have taken issue with the influence of state and corporate power over electric life and provide insight into necessary reform of the existing, privatized electricity system to serve the needs of its customer-constituents. In response to the climate crisis, some have argued that expanding deregulated electricity markets in the South, where most utilities remain vertically integrated monopolies, could enable more rapid build-out of utility-scale and distributed renewable energies.[20] Here, I urge a different pathway in asking how the investor-owned companies that assert themselves as "public" utilities could live up to their mandate.

Georgia Power is an existing planning authority that functions as an economic development agent and infrastructure provider with the consent of the Public Service Commission and funding collected from ratepayers' bills. The utility contributed to a regional economic development agenda that exerted a regional and national influence on energy and industrial policy by cultivating a "pro-business" climate characterized by anti-union politics, low taxes (and thus an underfunded, weak social safety net), and limited environmental regulation that kept the costs of doing business, including energy costs, low for industry. As Chandra Childers describes, this agenda "was never designed to help the vast majority of working Southerners; rather, it reflects efforts to ensure continued access to the cheap labor of Black people following emancipation."[21] While low-wage workers are increasingly diverse, overwhelmingly, Black and Brown workers, especially women, remain underpaid, devalued, and forced to manage the unaffordable costs of everyday life like the utility bill.

From this vantage point, the commodification of electric life is also an entry point to study the hegemonic power relations that define the US energy system and broader economy in which social reproduction—such as healthcare, housing, and childcare—is largely privatized. Advocates who have come before the Public Service Commission consistently probe the uses of state power to support the utility. Just as the speaker who opened this chapter explained, the burden of electricity bills is also about the value of the wage as well as the quality of housing that defines energy efficiency.[22] Contemporary energy politics unite the experiences of being cost-burdened as a customer and receiving low wages as a worker with the existential threat of climate change. The utility cannot address these intertwined issues on its own. However, as a regulated corporation, the utility has unique influence over our everyday and future lives, including the power to mitigate the theft of our future through climate catastrophe. I interrogate how this regulatory and material infrastructure could be transformed in the interest of a truly public power system. Asking how the utility could change drives at broader questions of how public resources are organized and how infrastructures could instead provide for affordable and clean power and the means of social reproduction as part of a universal social wage.

A social wage refers to provision of a basic standard of living that is not linked to wage labor, encompassing entitlements that derive from living within a society. In Atlanta today, the social wage includes public greenspaces, schools, libraries, water and road infrastructure, government-funded healthcare (albeit limited), among other municipal, state, and federal services and benefits. The social wage encompasses these and broader demands for social protection often referenced in terms of a social welfare state, such as public investment in affordable housing, subsidized transportation, and affordable energy, environmental quality, and protection from climate change. The social wage "decommodifies social protection," ensuring that individual and collective welfare is not dependent on the market.[23] The notion that affordable electricity and a clean environment should be part of a social wage form the basis of a democratic, everyday politics of utility reform. This case for reform offers a prefigurative politics—that is, political organizing that strives to build desired energy futures in the present—by asking how our collective investment in electricity infrastructure could be redirected to sustain and nourish collective social and environmental health.

Electric Life digs into these questions in Atlanta to investigate the long history of the city's electric utility from 1883, when Georgia Power's predecessor was chartered, to 2025. Braiding the histories of Georgia Power and social movements in Atlanta that have challenged the monopoly control of the private utility, I trace the long history of consumer, labor, and environmental organizing as antecedents to contemporary energy democracy organizing. This diffuse movement has worked through the legislature, courts, and Public Service Commission to influence the electric utility—seeking to promote collective oversight and ensure accountability to public demands so that ordinary people have greater control over the energy system they depend on. From 2018 to 2023, I participated in coalition organizing for energy democracy in Georgia and observed five central concerns emerge that I follow across the chapters in this book: representation in electricity governance for Atlantans, affordability of electricity bills, environmental justice and climate action, protecting workers' rights, and building pathways into quality jobs. Achieving energy democracy requires confronting how utilities influence these and other dimensions of electric life.

This book emerges from the specific regional context of the US South where intransigence and climate skepticism have stymied action on climate change.[24] In 2017, Atlanta became the first major city in the South to commit to procure 100 percent of its energy from renewable energy.[25] While a powerful voice urging climate action, Atlanta is limited in its ability to enforce or realize its own local commitments given Georgia Power's monopoly. At the same time, Georgia has become a national leader in clean energy, securing more than $22 billion in private investment in utilities, industrial facilities, and manufacturing of solar panels, batteries, and EVs between the 2022 adoption of the IRA and 2025, while also firing up Plant Vogtle, the first new nuclear plant to be permitted since the Three Mile Island accident in the 1970s.[26] Georgia Power is linked to a wider regional energy network through its parent company, Southern Company. Southern is a holding company that made $26.7 billion in operating revenue in 2024.[27] It operates Georgia Power, alongside Alabama Power, Mississippi Power, gas distribution utilities, and generating facilities across the US, serving more than nine million customers and businesses, including 2.7 million customers in Georgia alone. Southern and other vertically integrated utilities in the region, including Duke, Dominion, and NextEra, are among the largest and most profitable utilities in the US.[28] While this book tells the story of electricity in Atlanta in relationship to Georgia, the region's central role in a clean energy future has wider implications and offers broad lessons to understand the constraints and opportunities of improving our energy and social infrastructure.

A Brief History of Commodifying Electric Life

Telling the history of electrification from Atlanta offers a unique perspective on the rise of monopoly utilities and their enduring control over the electric grid and mechanisms of electricity governance. The dominance of monopoly utilities was neither natural nor predetermined but politically enabled through public institutions.

A robust literature on the early twentieth-century Southern economy documents how public officials and corporations collaborated after the Civil War to expand the role of the state to build the industrial base of what newspaper editor Henry Grady termed the "New South" and diversify the plantation economy built by enslaved people.[29] Regional boosters pursued commercial, financial, and industrial investment and worked with elected officials to reconfigure the state as "an agent of industrial prosperity" while maintaining a strict white supremacist racial hierarchy.[30] Clyde Woods studied the formation and reconstitution of political power by plantation owners and new industry as "regional blocs" that pioneered development agendas. In his analyses of the Mississippi Delta and New Orleans, Woods theorizes the regional bloc as a relatively stable political alliance of state and civil society actors spanning racial, ethnic, and class positions that achieves hegemony through state power and nonstate institutions.[31] In Atlanta, Clarence Stone similarly observes a consolidation of power following World War II through an interracial coalition of business, faith, and community leaders that he terms an "urban regime."[32] Subsequent histories of Atlanta have critiqued enduring racial inequality and the democratic deficit of the urban regime, even after the civil rights movements that inaugurated a *New* New South.[33] Georgia Power is anecdotally recognized in histories of the city and state as an "old-guard" member of the urban elite, yet the managing function of electric utilities within the quest for development is often overlooked.[34] This is despite what Manning Marable diagnoses as the Southern Company's integral role as "a forum for the coordination of electric power with the financial and industrial development of the entire region."[35] In the next chapter, I dig into the specific ways in which Georgia Power shaped urbanization in Atlanta and the tensions between the urban regime and the wider regional bloc coordinated through the state government that influenced Georgia's growth. Georgia Power's ascendence as a state actor curtailed the city's authority to advocate for its own energy future while the utility emerged as a planning entity that defined the regional trajectory of energy-intensive development. The quasi-public position of the utility as a regulated service provider ensured customers and revenue and the company's growth unfolded through the homes of ordinary ratepayers.

When the sale of electricity first began in the late nineteenth century, companies concentrated in urban areas sold production equipment for on-site generation. Improvements in steam turbines and the invention of alternating current allowed power companies to transmit power from a central power-generating station over longer distances, seeding the growth of the modern electric grid.[36] These plants could provide lower rates by spreading out the costs of construction and financing over a larger customer base. Yet, large, fixed capital investments only made sense if utilities had demand for their product. Initially, both private and public utilities competed for customers. In the first decade of the twentieth century, municipal power companies grew at double the rate of private firms as towns and cities consolidated basic service provision for water, transportation, and electricity.[37] Public power came to be overshadowed by private electric service under vertically integrated monopolies.

Competition between firms gave way to consolidation nationwide, following a trend toward monopolization across American industry.[38] This history is often told through the biography of Samuel Insull, who started in the electric industry working for Thomas Edison and pioneered organizational strategies to make electricity profitable. Where firms had initially charged flat rates per light bulb or socket installed, Insull sold current, adopting a metering system for volumetric rates billed in kilowatt-hours.[39] As president of Edison Chicago, Insull also recruited a diversity of customers with different power demands to maximize use of a power-generating station: Streetcar companies used the most power at dusk and dawn, municipal lighting systems and amusement parks at night, factories during the day, and homes in the morning and evening.[40] Achieving this optimum mix of consumers required control over a service territory. To preempt antitrust action that had led to the breakup of monopolies in other industries and quiet calls for public ownership of electric service, Insull used his position as president of the trade group the National Electric Light Association (NELA) to promote the idea that utilities are "natural monopolies" that should be regulated by the state. Regulators could oversee monopoly providers to prevent abuses while ensuing efficient customer service without duplicating infrastructure like transmission and distribution lines.[41] Far from being the thoughts of a lone figure, Insull's ideas gained purchase through utility executives across the country. As a member of the NELA Legislative Committee, Preston Arkwright, who would become the first president of the Georgia Power Company, championed regulation in the South. In 1907, Georgia was among the first states to adopt electricity regulation by the public service commission. By 1920, commissions governed electricity in thirty-five states, and today all states have such commissions.[42]

Public service commissions reorganized state capacity not to provide electric service like roads, schools, water, and other necessary public infrastructure but to oversee private power providers. Over time, these commissions have aided electricity capital and ensured its profitability. As Ruth Wilson Gilmore's theorizes, the organization of state capacities serves "in common sense and law, [to] ideologically bind state and society" to certain formations won by regional blocs.[43] This is exemplified by the utility consensus that united corporate and state interests.[44] Conor Harrison observes that granting a monopoly franchise to sell current to a captive customer base in a noncompetitive service territory created investment certainty for utilities "as state utility commissions essentially guaranteed profits for investors."[45] In turn, regulators review the rates and fees utilities charge to ensure they can recover their costs and garner an allowable rate of return without capturing monopoly profits. In his canonical text, *Principles of Public Utility Rates*, James Bonbright theorized that in the absence of a market in which customers could protest the high cost of energy by choosing to purchase power elsewhere, a regulator acts "as a substitute" for competition to determine the value of a utility's product.[46] The value that the regulator assigns in its determination of rates is not based on the cost of production alone but the value of service to customers. Rates are calculated from the capitalization of utility assets to allow utilities to recoup investment in power plants and transmission infrastructure.[47] An allowable rate of return, which customers also pay through their electricity bills, is distributed as dividends to shareholders to ensure utilities can continue to garner capital for investment.[48] Regulation thus provides security that the utility's product will sell with a guaranteed return to shareholders who earn a profit from electric life. This accounting design incentivized construction of new power plants rather than efficiency or maintenance and set in motion what Hirsh termed a "grow-and-build strategy," organized around growing demand for electricity and building infrastructure to meet that demand.[49] Utilities launched advertising campaigns to encourage home electrification and to recruit business to their service territories that would justify building new power plants—making them powerful boosters for regional growth.

The Great Depression threw the monopoly utility model into question. Rampant speculation on utility stock was blamed for the 1929 market crash, instigating federal action.[50] Hearings held by the Federal Trade Commission documented how utilities influenced and then exploited regulation, including in Georgia. Georgia Power President Arkwright detailed numerous strategies his company had used to curry favor with Public Service Commissioners and other civic leaders, including sharing material with trade groups and directly hiring or funding professors to design curriculum on utility operations and regulation for economists, engineers, and lawyers. Arkwright compared his efforts to review curriculum and "in some cases . . . [have] text books changed or corrected in the making" to efforts by Southerners "to examine the history text books. . . . to see that incorrect representation and injustice should not be done to the Confederate cause and taught to their own children."[51] In equating propaganda efforts of the Lost Cause with evangelizing for the investor-owned utility model, Arkwright aligned the public mandate of the utility with a white supremacist vision for development of the New South. Federal oversight imposed through the 1934 establishment of the Securities and Exchange Commission and the Public Utility Holding Company Act of 1935 ultimately restricted speculative utility practices, including utilities owning companies that were not directly tied to the electricity business and the combination of utilities into holding companies that were not contiguous.[52] These reforms curtailed growth strategies and pushed utilities to focus on development within their service territories, yet saw limited success in breaking up corporate control of the electric industry.

The federal government also promoted public power as an alternative to investor-owned utilities. The Tennessee Valley Authority (TVA) was established north of Georgia Power's territory in 1933. President Roosevelt envisioned TVA as a "yardstick" to measure private utility company service.[53] Through executive order, Roosevelt also created the Rural Electrification Administration.[54] In his comparative analysis of TVA and Georgia Power, Jason Cebul argues that the political popularity of TVA—which launched local planning and agricultural modernization projects and sold cheap hydroelectric power without a cut going to shareholders—chastened private utilities who borrowed the tactics of their public competitors to expand electrification. Wendell Wilkie, CEO of the Commonwealth & Southern Corporation, a holding company that had acquired Georgia Power, asserted that the "fairest answer and the surest defense against competitive government power projects . . . are for privately operated power systems to so increase their household use of electricity that domestic rates may be reduced to the point where the differential between the cost of service for private companies and government operations becomes so inconsequential that it will no longer constitute a political factor."[55] In other words, expanding residential electric intensity offered a new way for Georgia Power to seize economies of scale, lower their cost per kilowatt-hour, and win popular political support.

This ambition to expand electric life was aided by other New Deal programs. As Matthew Huber observes in his analysis of oil politics, the material reorganization of "reproductive geographies" through New Deal housing, labor, infrastructure, and environmental policy restructured labor contracts and secured lending and private insurance markets to expand an energy-intensive way of life.[56] These measures induced private provision of consumer goods, including housing, appliances, and farm equipment all wired into the electric grid.[57] Georgia Power would fight TVA in court and won a lawsuit to erect a fence around the TVA to prevent its expansion. Despite this, into the 1960s, Georgia Power cited "its principal public relations problem [as] the issue of Federalized power."[58] The "TVA battle scars" reminded the company to maintain "skepticism" and an "unwavering watch for signs of future federal or state intrusion" while also working to manage "public perception" of the private monopoly model.[59] It actively opposed public power alternatives even as it leaned on public authority and funding to expand the intimate reaches of electricity.[60]

Historical analyses of the frayed collective project of the New Deal reveal its limits to achieve energy democracy. This important expansion of the social welfare state included not only energy policy but also Social Security, a framework for labor law, and other investments in social and physical infrastructure that have continued to inspire social democratic platforms, including the Green New Deal to address climate change.[61] Yet, instead of a comprehensive social wage that decommodified basic goods and services necessary for social reproduction adopted in other parts of the world, for Huber, New Deal consumer democracy laid the groundwork for an individualistic and entrepreneurial politics of neoliberalism.[62] Nancy MacLean similarly studies how New Deal programs were reconfigured through regional racial prejudices to evade integration.[63] The Southern Democrat congressional delegation resisted federal policy that would threaten the regional labor regime where the "psychological wage" of whiteness, racial terror, and Jim Crow sustained a segregated, low-wage economy.[64] Locally administered and means-tested programs supplanted universal and federal ones. New Deal programs and wartime federal investments that followed allowed regional elite to augment their "power by developing the regional economy without democratizing the polity."[65] The racist and regional roots of Southern developmentalism can thus be seen as one origin point for the rise of neoliberalism in the 1970s and 1980s, which Jamie Peck and Nik Theodore define as a "project of state-facilitated market rule,"[66] characterized by deregulation and austerity that divested from state social services and increased labor precarity in the interest of marketization. With the dismantling of the already-limited New Deal institutions for social reproduction, "neoliberals persistently exhort individuals [and their families] to take responsibility for their own fate."[67] This is not a new idea but what Martin Luther King Jr. described as a belief advocated by "white Americans" that Black people should lift themselves up by their own bootstraps while refusing to recognize how many "have been left bootless as a result of all of these years of oppression and as a result of a society that deliberately made his color a stigma and something worthless and degrading."[68]

I am less focused on the periodization of neoliberalism than tracing how state capacities to sustain "industrial prosperity" and enable corporate power have been constructed over decades in ways that accept and maintain racialized poverty. For instance, in the case of one Georgia Public Service Commissioner, the high cost of new generation assets that would increase power bills was downplayed, given their "outsized effect on economic development."[69] Or, as an energy advocate translated, "keeping electricity prices low is not a priority for the PSC. Attracting business is because people need jobs."[70] Instead of provisioning for social reproduction, collective interest is framed around creating economic opportunity regardless of the uneven distribution of the benefits and burdens of growth.

When utility reform was again broached in the 1970s to address the growing unaffordability of power provided through the utility consensus, states began to consider deregulation to introduce wholesale and retail power markets. This push toward deregulation that became a hallmark of the ascendent neoliberal order was ultimately forestalled in the South by regulated utilities who used their political influence to prevent restructuring. Conor Harrison and Shelley Welton argue utilities realized electricity markets threatened their consistent rates of return. They used their close relationship with state actors, including regulators and legislators, and lobbied their largest industrial customers who would stand to benefit from electricity markets, to dissuade from deregulation.[71] Public power, electricity markets, and regulated, monopoly utilities all rely to some degree on state power. What is important to understand from this history is that the utility industry in the South thrives not as an independent market actor but in using its quasi-public position and regulatory certainty to coordinate and encourage growth.

The mobilization of state capacities in and beyond the regulation of electricity had pronounced national and global effects in driving down the value of labor.[72] Barry Bluestone and Bennett Harrison interrogate how southern states actively promoted regional competition and cultivated a pro-business climate "ranked in terms of low labor costs, low tax rates, [and] low energy costs."[73] Implicit in this race to the bottom in inter-regional competition is under-investment in social reproduction. In vacating a role for the state as provider of social welfare, individuals, families, and religious institutions are left responsible for the reproduction of social life, even as provisioning for these networks of daily life justify Georgia Power's existence.

Energy democracy must grapple with the evolution of public authority to uphold private power providers that was simultaneously organized around and through the home. Electricity capital profits from our quotidian demand for power in electric life. However, far from recognizing ratepayers' contributions to this shared electric infrastructure, the utility consensus around private power provision effaces these social relationships in the interest of a more narrow vision of economic growth that has accepted and promoted racialized uneven development in Atlanta. To understand how this system came to be requires diving deeper into the question of how electricity is valued.

Electricity Capital and the Time of Electric Life

In its 2017 annual report, titled *Imagine*, Southern Company invited its reader to "Imagine, if you will, a company that is building the future of energy."[74] Vibrant photographs detail the ways in which this future-oriented imagination shapes the lives of every person in the company's service territory. Electricity drives whirring medical equipment keeping infants alive in hospitals, irrigation on farms, and the scoreboard at the Atlanta United soccer games. The report highlights the diversity of the company's twenty-eight thousand employees, from linemen to customer service representatives, researchers, and accountants. The embodied, if not familial, narrative disappears when the introduction concludes with a provocation to "Imagine creating value." Devoid of the customers, workers, and users who enliven the grid, the report goes on to explain that Southern Company's "commitment to keep customers at the center of all we do . . . ultimately translates to value creation for investors," measured in dividends paid every quarter for shareholders since 1948.[75] For Southern Company, value is equivalent to its stock price—what shareholders are willing to pay for a piece of the company. Of course, this price is determined not in the market alone but also by regulation and the guarantee of an established rate of return that shareholders can expect on their investment.

The company's stated commitment to finding value in its customers offers insight into how the networks of electric life uphold and reproduce electricity capital. As people use and pay for electricity, Southern Company and its shareholders (not to mention the thousands of other companies who depend on its product) earn a profit, including from those who can least afford it. Since the financial crisis, a growing chorus of scholars have critiqued weakening regulation and increasing monopolization across the economy that allows shareholders to extract rent from their control over scarce resources, leading to increasing income inequality. In the utility industry this is not a new concern. Labor unions and consumers protested the formation of utility monopolies as a tax on working people from their origin.[76] Yet, in her genealogy of value, Mariana Mazzucato problematizes that such critiques "do not fundamentally challenge how value extraction occurs" and persists, which requires instead the need to consider "where value comes from in the first place."[77] From this perspective, the price something can garner in the market does not tell the whole story. Neither does the distribution of revenue—shareholders earn more than is fair, while customers pay too much—but rather, the way we define value is missing something altogether. So, how should we define value? And what role does electricity have in this process?

Over the last fifteen years, critical energy scholars have argued that to answer these questions requires going back to the Industrial Revolution when the uses of fossil fuel energy began to increase at the same time that many more people began to work for wages.[78] Classical political economists writing at this time, such as Adam Smith and David Ricardo, first offered a definition of value based in labor, or labor-time. In his history of the ascendence of steam power in the British Industrial Revolution, Andreas Malm demonstrates how fossil fuels integrate into value creation. He observes that fossil fuels transform our daily lives to such a degree that they change our quotidian and historical sense of time.[79] He turns to Moishe Postone, who illustrates that during the Industrial Revolution, the time of seasons and the diurnal rhythms of sunup and sundown were replaced with the proliferation of clocks as time became a measurable thing able to be "spent" in which "motion, events, and action occur. . . . [and become] divisible into equal, constant, and nonqualitative units."[80] Gas lamps that extended the working day and coal-powered machines that kept workers on pace in new factories created the conditions for wage labor as the day was divided into working hours. Fossil fuels disciplined workers to the time clock and the pressures of wage labor in which the threat of losing one's job was ever present as coal also provided a means to increase productivity by powering machines that could be substituted for workers. Thus, Malm argues fossil fuels became a "general lever for surplus-value production."[81] His analysis builds on Marx's critique of classical political economy's labor theory of value. Marx argued that there is no innate or natural measure of value, but rather value is a socially contingent category specific to capitalism.[82] Labor-time is thus the source of value within capitalism because the universality of waged work made commodities produced by workers under given technical and social conditions comparable.

Within this definition of value there is an important temporal dynamic at play. Marx theorized that labor is a special commodity because workers can produce more value than the cost of goods needed to reproduce themselves and their families. Employers can pay workers the minimum amount needed to survive and pocket the extra, or surplus, value. Thus, value is extracted from workers in the production process. However, this exploitation is not obvious because value is realized in the market when a commodity is sold. There, a commodity takes on an independent character separate from the work hours of people who made it. Nevertheless, value cannot be reduced to a monetary measure of price. It describes the social relations embedded in wage labor and commodity production that shape our experience as workers and consumers.[83] For Postone value is thus a measure of social wealth based on socially necessary labor-time that comes to dominate our social and economic lives. Value is maintained as a social idea because people's time is structured through commodity relations, shaping how much our labor is worth, what we can afford to buy, how much we work, how we judge others, and how we define our own sense of personal worth, ambitions, and sense of progress.[84] Thus, in going to work, workers are not only producing light bulbs, electricity, or any other good or service for sale (in addition to profit for their employer) but also the social relations based on value in which (for most people in Atlanta) wage labor is required to earn money to pay for rent, food, utilities, and everything else.

Many scholars have argued that the central emphasis on wage labor in labor theories value is overly simplified. Stuart Hall asserts that in practice, value "operates through and because of the culturally specific character of labour power, rather than—as the classical theory would have us believe—by systematically eroding those distinctions."[85] In other words, across the economy and within workplaces, racial and gender discrimination mean women and nonwhite people receive lower pay, sometimes for the same work, such that their identity become another means to extract value. Beverley Mullings goes further, to argue that the primacy given to the wage relation as a source of value also omits "the informal, part-time, temporary, casualized, unpaid and even unfree forms of work" coerced within capitalism and colonialism.[86] She points to Black, Indigenous, and feminist scholars who document the ways in which racialization was used to steal land, devalue labor-time, and coerce labor under conditions of slavery, sharecropping, convict leasing, and other, persistent forms of unfree labor.[87] In particular, Black feminist scholars document how the plantation economy that provided material inputs for industrial production relied on the forced labor and social reproductive energies of enslaved women who worked in fields, birthed children, and cared for families. Racism and sexism devalued Black women's paid labor after the abolition of slavery in defining the kinds of work, education, and wages women were afforded.[88] Across the economy, jobs seen as "women's work" remain underpaid as waged work or unpaid within the home where caring for children, cooking, and cleaning are necessary to reproduce the next generation of workers and replenish workers so they can return to work the next day, and yet are not recognized as valuable, or worthy of a wage. Feminist scholars argue that employer's reliance on this labor of social reproduction represents an on-going taking or "primitive accumulation" of care given out of love or obligation.[89] As Nancy Fraser aptly summarizes, capitalism is often defined as "an economic system based on private property and market exchange, wage labor and production for profit," but it can be more accurately understood as "a societal order that empowers a profit-driven economy to prey on the extra-economic supports it needs to function—wealth expropriated from nature and subject peoples; multiple forms of carework, chronically undervalued when not wholly disavowed; public goods and public powers, which capital both requires and tries to curtail; the energy and creativity of working people."[90] Bringing together feminist studies of social reproduction and Black radical critiques of racial capitalism to understand the conditions of unfreedom, devaluation, and obligation hidden in the value form, we can see how profit ensnares both the time of wage labor and life times beyond the wage. As Myles Lennon observes, we cannot theorize energy outside these processes as the logic of industrialized energy was "traditionally rooted in an anthropocentric white supremacy that debased matter, nature, and nonwhite bodies in intersecting ways."[91]

Working from this understanding of value as a social relation broadly based on control over people's time offers a new vantage point into how electricity both functions within and fuels the inherent instability of capital's growth mandate. As employers compete, they seek ways to drive down costs to capture value, such as swapping labor-saving technologies to increase worker productivity or relocating to regions with lower labor cost to save on their total wage bill. Increases in labor productivity accelerate a "treadmill" dynamic as capturing value requires new ways to invest capital and new ways to usefully re-employ people who remain the source of value. Technological improvements, and social gains won through popular organizing, create a chronological sense of progress,[92] yet, over time, progress has also demanded new resources and greater inputs of energy, including electricity. Energy historian Vaclav Smil argues the social and technological advancements of "modern industrial and post-industrial societies" would not have been possible without "the conversion of an increasingly higher share of fossil fuels into electricity, the most flexible and convenient form of energy."[93]

Electricity reorganized labor relations from refrigeration that changed food processing to automation of assembly lines and checkout systems. Countless other labor-eliminating, time-saving innovations opened new circuits of employment and powered the expanded reproduction of capital across urban space.[94] New forms of leisure, including amusement park rides, department stores, radio, and movies offered electricity-intensive "purchases that were predictable and endlessly repeatable," erasing "the line between commodities and experience."[95] The rise of advertising, information, digital communication, and messaging brought to users, consumers, and viewers through electrified TVs, billboards, phones, and computers create new forms of labor with the growth of the service sector just as they enroll people's time in new ways to manufacture desire, cultivate brand loyalty, and reproduce consumer subjectivities. As David Harvey describes, consumer demands draw new moments of life into the reach of capital and hold time "captive to capital's struggle to absorb the necessity of its permanent compound growth."[96] The improved efficiency of production does not mean workers benefit from more free time for leisure. Instead, near-uninterrupted access to electricity at the flick of a switch in much of the US makes a twenty-four seven economy possible.[97] The time of work stretches through the day and night, while time away from work becomes commodified.[98] Apps and platforms that "buy and sell our attention" and data exert control over time.[99]

A wealth of historical research interrogates how electrification transformed the home and social reproduction.[100] Mass consumption of durable household goods such as washing machines, ovens, refrigerators, and—of course—electricity to power them revolutionized waged and unwaged housework. Improved household efficiencies "freed" women to seek paid employment and reorganized paid domestic labor, such as when Ruth Cowan observes, "the Jones's down the block decided to junk their washtub and buy a washing machine, and the Smiths around the corner fired the maid and bought a vacuum cleaner."[101] In her study of the industrialization of women's work, Susan Strasser complicates a received division of labor between productive or value-generating waged work on the job and the often-unwaged or underpaid work in the home seen "to be without value."[102] The unwaged work of social reproduction is not recognized for its economic contribution within most calculations of economic activity,[103] yet is integral to any utility's business model. As Wendell Wilkie claimed in the 1930s, household consumption of electricity drove new business opportunities for and political acceptance of the privatized utility model. Home electrification opens new circuits through which, in the words of Gargi Bhattacharyya, "non-work lives increasingly come to service capital."[104] Rather than seeing work and home as distinct spheres, Katharyne Mitchell, Sallie Marston, and Cindi Katz note that electrified communication and information technologies collapse the boundaries between production and reproduction such that people "[are] permanently mobilized" within "new technologies of power."[105] The flexibilization of labor pushes into the time of life. Melinda Cooper expounds that just-in-time schedules, night work, and overtime blur home and work, "destandardizing labor itself."[106] Across these dimensions, electricity capital accrues in the transformation of social life where electricity is used.

To understand how capital captures time, Beverley Mullings and Neferti Tadiar think through the myriad kinds of work required in "life-work" or "life-time." Tadiar proposes the concept of "life-time" as a corollary to labor-time. Life-time strives to account more accurately for the diverse kinds of unconventional and underpaid forms of work as well as social conventions, cooperation, and experiences, "which appear to lie outside the formal sites of labor exploitation and yet are continuing sources of appropriable value."[107] This is seen in the care work most often undertaken by women to budget, curtail, and manage household energy use to live within a month's balance sheet; the embodied physical experience of living in a home that is too hot or cold; or the time on the phone or in community action agency offices to seek out public and charitable assistance for energy bills.[108] Beyond struggling to pay for electricity, countless other unpaid (or underpaid) aspects of social reproduction subsidize the utility: the work of caring for family members suffering from asthma linked to coal-fired power pollution, training to prepare oneself for work, or advocacy of unions that garner political support for the utility's business model. The commodification of everything—including electricity—exerts pressure on daily life through broad infrastructures of financial, fiscal, institutional, regulatory, legislative, and ideological authority. Value circulates as capital enrolls people's time at and beyond the jobsite with electricity as a connective tissue between the blurred pressures of the time clock, media's competition for our attention and commodification of data on how our time is spent, and the gendered expropriation of unpaid labor, among other financialized and precaritized social reproductive relations.

The power of capital to control time reaches its apex in its measured disregard for the existential threat of climate change. Electricity capital not only reconfigures time in the present but steals future time in the interest of maintaining profit from operating infrastructures that power carbon-intensive electric life. The quest to capture value wreaks havoc on ecologies in which production is embedded, "expropriating violence on collective life itself."[109] The effects of violence, which Gilmore defines as "the cause of premature deaths," can accumulate slowly in everyday exposure to environmental pollutants that contribute to respiratory disease and cancers or in the gradual buildup of atmospheric greenhouse gas emissions.[110] These background environmental changes compound and accelerate the reoccurrence of five-hundred-year weather events as sudden, wrathful episodes become more frequent, from hurricanes that pummel Georgia's coast to floods that transform creeks into rivers and ravage Atlanta neighborhoods. Vulnerabilities to climate change, environmental injustice, and premature death amass along the lines of power and privilege that shape access to work, wealth, property, and the ability to capture surplus value or be extracted from. Control over one's time, which Gilmore asserts is "*the* resource of life," is political.[111]

Rather than seeking a general theory of equivalence akin to Marx's standardized measure of socially necessary labor-time that made work under given technological conditions comparable, I affirm a long tradition of feminist scholarship that observes this measure of value fails to account for the socially necessary practices of social reproduction required to make life and sustain capital. Even as electricity and the technologies it powers blur the space between labor-time and life-time, to collapse these categories within a unitary definition of social reproduction risks overgeneralization. However, thinking through the transformation of electric life reveals the iterative construction of hegemonic labor-energy regime that shapes thousands of daily choices by electricity customers. Quotidian uses of electricity are the basis of the utility's social license to operate. The vital need for power is why the state regulates electricity and authorizes a generous rate of return. Studying electric life as the terrain of electricity use in everyday life provides a lens to see how time is exploited by the utility and brings together the "fragmentary and disconnected" experience of workers and consumers.[112] To recognize people's quotidian needs for energy as the utility's source of value reverses the power dynamics embedded in the electric grid. Ultimately, value creation for shareholders depends on the needs and energy uses of people in the service territory that the utility controls. This begs the question as to how our collective time and resources could better serve our communities. Energy democracy invites popular participation to answer this question as to how the electric utility, and the regulatory system that sustains it, could better support electric life and the broader needs of social reproduction.

Feminist Urban Political Ecology for Energy Democracy

Imaginaries of building an energy system responsive to workers, energy insecure households, and frontline and fenceline communities motivate organizing for a just and equitable energy transition. The term "energy democracy" stems from the German climate justice movement and has since been taken up by environmental justice organizations, labor unions, and community and consumer advocacy groups.[113] Energy democracy asserts that the transition from fossil fuels to renewable energy resources is an opening to rebuild the energy system and "literally and figuratively" redistribute power to workers, Indigenous people, environmental justice communities, and others marginalized within existing infrastructures.[114]

Despite these emancipatory aims, in practice, a broad constellation of articulated social movements have claimed the mantle of energy democracy with different policy and governance goals.[115] In the US, campaigns have concentrated on local control for public or community ownership of distributed energy resources through municipalizing electric companies, establishing consumer choice aggregation, or promoting inclusive solar financing.[116] Social scientists and legal scholars caution against technological determinism that equates distributed renewable energy systems with democratic ones, noting how such policies can fall short in expanding the polity if they remain focused on empowering propertied citizen-consumers to expand participation through new market mechanisms for net-metering or demand response,[117] not least because solar adoption skews toward whiter and wealthier households, patterning inequalities seen in property ownership.[118] In contrast, policies focused on addressing racial and economic disparities in energy access have contested how power is distributed through the energy system. Community solar and tribal energy projects assert collective ownership over renewable energy resources to provide access for renters and low-income homeowners or build energy sovereignty.[119] In parallel, efforts to empower energy insecure households to participate in regulation attempt to change the terms of regulation at public service commissions to advance energy justice.[120] These mechanisms of reform are prefigurative politics that work to build just energy futures in the present by showing how the physical, financial, and political relations of energy can be transformed to support life.[121]

Energy democracy envisions how the electric grid as a collective project could sustain community and household reproduction. Following James Angel, I assert that democratizing energy must entail a politics of social reproduction that begins with asking how and why people need and use energy and what systems can best meet these collective needs.[122] Centering people's right to participation and self-determination in the energy transition builds on the long tradition of the grassroots environmental justice movement that, among other principles, promotes "economic alternatives" to respect people and their connection to the natural world.[123] This is also aligned with more recent calls for energy justice that interrogate how institutions, regulatory processes, and material conditions of energy production, distribution, and use must change to achieve a just energy system. Yet, energy democracy is not only about addressing inequality in the process of governance to accommodate more participation or greater responsiveness to advocates who are already engaged in energy politics. Rather, it is about people whose lifetimes are unrecognized in energy regulation making and realizing political claims. Energy democracy contests the social conditions that allow energy poverty, environmental injustice, and the expropriation of nature, public resources, and racialized and gendered labor.[124] It seeks to end the racialized and gendered hierarchies of valuation/devaluation that reproduce racialized uneven development through energy infrastructures. Thus, in politicizing the energy transition, energy democracy rejects control by incumbent operators and the framework of shareholder ownership to argue for reform or wholesale system change to achieve a democratic, socially just, and less ecologically damaging energy system in which ordinary people can decide "who owns and operates our energy systems, how energy is produced, and for what purpose."[125] Organizing demands that ratepayers who sustain the energy system participate in governance to design the collective life of electricity. From this perspective, energy democracy is about much more than decarbonization and asks how efforts to reimagine public infrastructures can contribute to (re)building a social wage, including universal access to affordable and clean energy.

Environmental justice and urban political ecology scholars have studied the creation and transformation of material, biophysical natures within and through urban environments as well as how both access to resources and ideas about nature are mobilized to justify enduring inequalities.[126] I adopt a *feminist* approach to urban political ecology that builds on the traditions of feminist political ecology, which calls for postcolonial, intersectional, and embodied analyses to analyze how gender, race, class, nationality, and ability produce and are reproduced through urban environments.[127] This framework integrates feminist political economy's study of social reproduction to assess how life is materially constituted within urban landscapes and how the utility harnesses the time of electric life across mundane infrastructures.[128] Within the patterns of everyday routines also emerge the possibility of radically different ways of organizing electricity that challenge the use of state power to advance accumulation.

Energy democracy organizers draw from intergenerational traditions of the environmental justice, labor, and civil rights movements to reject a racial capitalist vision of development predicated on control over people's time. At various points, energy advocates have sought inspiration from Ella Baker's life's work.[129] In her biography of Ella Baker, Barbara Ransby describes Baker's practice of radical democracy that responded to socially and historically specific "parameters of exclusion" defined through race, gender, and class. Disrupting the "cold impersonalism of the market-driven dominant culture," Baker observed that institutions are comprised of people who can choose to enact to create a "more human social order."[130] Energy democracy gives shape to these lived politics to ask how the systems that govern energy can be reorganized to use time and resources to support, sustain, and nourish communities now and in the future.[131] In Atlanta, faith leaders, union members, environmentalists, and Georgia Power customers link the cost of energy to a wider urban politics concerned with low-paying jobs, exorbitant housing costs, gentrification, policing, and pollution as part of a right to urban life.[132] My account of energy democracy organizing draws on scholars of environmental and energy politics in Atlanta who have long documented these connections, including William Bryan, Casey Cater, Christopher Manganiello, Richard Milligan, Na'Taki Osborne Jelks, Kristin Phillips, and Christopher Sellers.[133]

This book bridges archival and ethnographic research to investigate the *long durée* of the city's electricity politics and illuminate how historical relationships influence the present. I conducted more than ninety interviews with solar installers, city workers, union members, trainers, and environmental advocates working in Georgia and nationally.[134] Georgia Power officials did not respond to requests for interviews.[135] Instead, I compile my understanding of the company from analysis of news articles, press releases, annual reports, and official testimony in recent and historical regulatory proceedings as well as archival data from the Atlanta History Center, the Southern Labor Archives, the Richard B. Russell Library, and the Georgia State Archives. From 2018 to 2023, I attended, and at times participated in, city council meetings, hearings before the Public Service Commission, workshops, industry conferences, and meetings of environmental and energy justice organizations. These ranged from majority-white and volunteer environmental campaigns for 100 percent clean energy with the Sierra Club, trainings on integrated resource planning hosted by the nonprofit Southface, energy justice forums convened by the Partnership for Southern Equity, Zoom meetings of the Just Transition for Georgia study group organized to support a Green New Deal, and the grassroots project We the Plug Tho that formed to protest utility-shutoffs and lead mutual aid campaigns during the COVID-19 pandemic. From my vantage point in Athens, Georgia, where I did my PhD, and later Knoxville, Tennessee, where I now live and work, I attended virtual meetings and drove into town for events. My ethnographic engagement was limited by my distance from the city. Yet, following a tradition of engaged research, I contributed in small ways to some of the campaigns in this book, researching policy questions, making social media posts, and phone-banking for events.[136]

My research questions emerged in solidarity with energy democracy organizing. As Richa Nagar defines, building "situated solidarities" through ethnographic research is "inseparable from a deep commitment to ever-evolving critique that is grounded in the historical, geographical, and political contingencies of a given struggle."[137] Despite this call to introspection on the spaces I witnessed and contributed to throughout this research, part of my process of reflection in writing this book was a decision to spend less time studying the social movements that I learned from than the institutions they sought to change. Throughout the chapters, I reconstruct vignettes from fieldnotes or archival materials and interviews that demonstrate tactics organizations adopted to shift policy, win recognition of and redress for energy poverty, or campaign for clean energy. Some of these campaigns were successful; many were not. Throughout five years of participation in social movement organizing as an allied academic, I also witnessed interpersonal conflicts and systemic power imbalances that eroded trust in organizing projects until accumulated harms led to organizers burning out or groups dismantling altogether. At times movement spaces reproduced racism, sexism, homophobia, classism, and agism that we sought to change at Georgia Power and the Commission. My choice here to focus on the utility is not to condone or paper over the fractured and messy processes of organizing and coalition building. Rather, I seek to identify where and how points of convergence emerged to forge solidarity. In following the infrastructures of electricity to find these connections between different people, sites, and moments of organizing, we may, as Katherine McKittrick identifies, "illustrate the ways in which human geographies are, as a result of connections, made alterable."[138]

This engaged research praxis shapes my analysis of electric life and how everyday interactions with the electric grid influence organizing for change. My method to follow the times of electric life emerged from a workshop I attended with Georgia STAND-UP, a "think and act tank" for working communities. Each Thursday in the fall of 2019, a small group would meet at the International Brotherhood of Electrical Workers (IBEW) union hall to learn organizing strategies and discuss challenges we had faced within our own organizations. At one, Georgia State Professor Emeritus Harvey Newman offered a few lessons on his political experience within the regime politics that shape Atlanta. While some embrace Atlanta's identity as "the city too busy to hate," he shared that he had always found the words of civil rights leader Juanita Abernathy to be apt when she professed, "If Atlanta is the city too busy to hate, I'd hate to be there when they find the time." I thought often about Mrs. Abernathy's turn of phrase in writing, considering how electricity shapes our quotidian understandings of time. The calculation of kilowatt-hours, the economization of everyday life, and culturally embedded, value-laden ideas of time "well spent" cohere in small ways to uphold the utility and blame working-class people who struggle to afford commodified current for personal, if not moral, failures. My methodology was based on "finding the time" of electric life invisibilized, captured, devalued, or appropriated in the service of economic growth to recuperate and honor the everyday moments of survival, resistance, care, and joy against the appropriation of time. This account of the corporeal, emotional, and cooperative labors caught up in electric life defies measurement. While committed to reparative efforts to redress past and on-going harms, the goal of this book is not to measure the immense and incalculable moments across which people's time is and was devalued and stolen. Rather, I ask how a system based on value that diminishes people's humanity ever came to dominate our lives and energy system and what it would mean to change the investor-owned utility to uphold a *public* mandate to recognize and mitigate the pressures that electricity capital exerts on the everyday lives of customers and build life-affirming institutions.

Outline of the Book

In the next chapter, I turn to the urban politics of electric utilities to investigate how the utility consensus between Georgia Power and the state Public Service Commission both transformed and disempowered the city of Atlanta. I deconstruct Georgia Power's assertion that it is a public utility with a natural monopoly over Atlanta's electrical production. Following Casey Cater, I document how this monopoly emerged from the conjuncture of white anger, antitrust populism, and electric industry boosterism in the period between the 1906 Atlanta race massacre and World War II.[139] Evolving as a quasi-public entity and booster for the state, the utility promoted economic development that has not benefitted Atlantans equally as electric infrastructure reinforced classed and racialized segregation in Atlanta. Municipal efforts to address the racialized uneven development of the city, including racial disparities in energy bills, and provide for social reproduction are constrained by state preemption of city control over the electric grid that impedes local organizing to mitigate energy burden and climate change.

The chapter concludes with a discussion of Black Atlanta voters' 2020 lawsuit to institute a new method of electing Public Service Commissioners.

Chapter 2 scrutinizes the process of rate-making that ensures utility profits, even at the expense of customers who struggle to afford the conveniences of electric life. I document utility advertising to expand household consumption and how regulation by the Public Service Commission incentivized this grow-and-build model that allowed the utility to bundle the everyday demands for power in electric life as a financial asset to borrow capital for continued construction. Public campaigns beginning in the 1970s contested the utility's growth mandate. This chapter traces three consumer advocacy movements to lower the cost of electricity, change the practice of assessing fees, and fight a hike in mandatory fees.

Building on my analysis of the institutional and financial organization of utility ratemaking, chapter 3 asks how narratives of development wed the company to fossil fuels and authorized on-going harms to human and environmental health. It documents the political ecologies of Atlanta that harnessed water and coal resources for urban growth and the utility's subsequent actions to fight federal regulation of coal pollution under the Clean Air Act, culminating in continued climate change denial and delay. I consider the challenges of building common cause around localized environmental effects of electricity generation in reviewing missed opportunities for solidarity organizing between conservative, white rural residents and multiracial, environmental groups centered in Atlanta working to clean up coal ash. I contrast efforts to clean up the legacy waste of coal with successful organizing by environmentalists and conservatives who united within the Green Tea Party to advance solar at the Commission. Reflecting on this history, this chapter concludes in asking how a pilot, utility-managed distributed energy program could advance energy democracy.

In chapter 4, I shift attention to the contradictory role of the utility in the anti-union politics of the South. This chapter traces the history and function of unionization at the utility. It documents how workers contested the social organization of time through unions and civil rights lawsuits to improve working conditions. I contrast diverging labor regimes in the energy sector between the rise of contracting out for solar construction to nonunion firms and the unionized construction of Plant Vogtle. Across these sites, I argue worker organizing to support social reproduction must be seen as part of energy democracy organizing for a just transition.

Beyond the union apprenticeship, the utility utilizes other training programs to reproduce workers employed in the production of electric life. In chapter 5, I consider the historic and on-going role training plays in clean energy policy to create and manage a workforce for the utility and related energy industries. I trace a genealogy of job training through public policies to manage surplus labor from New Deal public works programs to the rise of neoliberal workfare. Against this historical context, I evaluate green jobs training programs that emerged in Atlanta during the 2008 financial crisis to prepare most-often young and un(der)employed people to the temporalities of wage labor and introduce skills needed to carry out green jobs. I observe how training becomes another site in which social reproductive labor subsidizes electric life based on a survey of green jobs training programs in Atlanta.

Electric Life concludes with a reflection on democratic organizing around the energy system to support collective needs. In observing the everyday moments of electric life, I learned that if power is maintained in part through consent to social, political, and economic relations, then power also resides within the daily decisions to reject or resist the appropriation of time.[140] I offer summary thoughts for utility reform based on these historical campaigns for a right to affordable energy as a basic need in a call to democratize and decommodify electric life.

1 Lighting the Atlanta Way

It was a quiet August evening when I pulled up to the Emmanuel Lutheran Church in Cascade Heights. A small yard sign advertised that the church was hosting a "Community Conversation on Clean Energy." It was the last town hall that the Mayor's Office of Resilience had scheduled to gather input from residents and workshop recommendations for how to implement the city's 100 percent clean energy resolution adopted the year prior.[1] Atlanta was the first major city in the US South to make such a commitment to procure all electricity from renewable sources by 2035. More than one hundred cities across the country adopted similar targets to continue action in line with the Paris Agreement after President Trump withdrew the US in 2017.[2] Given decades of federal inaction, city and state governments have historically taken leadership to mitigate climate change where possible and improve the resilience of urban infrastructure to its possible effects.[3] Despite this, Atlanta was still an outlier. In the South, few cities were willing to sign on to commitments that challenged powerful yet apathetic utilities and hostile state politicians who continue to question climate science or stay silent on the issue altogether.[4] Debate over how to implement Atlanta's 100 percent clean energy initiative encapsulates the constraints city government faces in contesting the utility that operates within the urban regime and at the state level as part of a wider regional bloc to set the pace of the energy transition and control development.

In the community center beside the church, I helped Marcus, a Sierra Club organizer, set up a buffet with trays of Jamaican goat curry. About forty people slowly trickled in, with advocates from a range of environmental, faith, and community organizations outnumbering the neighborhood residents present. People found their way to the food and piled rice and curry onto paper plates. The meeting began with a welcome from the pastor and a short presentation from the Mayor's office outlining the city's proposed actions to reach 100 percent clean energy. Discussion concentrated on how the city could achieve its targets if Georgia Power continued to slow-walk, if not directly stymie, the city's clean energy ambitions and whether it could implement two priorities that surfaced through public workshops: economic development and an equitable clean energy transition.[5] These twin priorities were intended to serve the needs of lower-income neighborhoods, like Cascade, where energy bills are an unaffordable expense that threaten residents' ability to remain in the gentrifying city.

The presentation was met with skepticism. A neighborhood resident asked why city resources were being used for energy planning at all because "you can say, we're doing this on Cascade, but you done gentrified Cascade—Cascade, ain't the same . . . community no more." Clean energy was a second-order priority divorced from economic precarity for Black residents in the neighborhood. A community organizer echoed his concern and asked for more specifics on how the plan would "help provide some relief for the six highest burdened zip codes in Atlanta?" She explained, "I see the word equity, but I didn't see any solutions or any movements that move towards residents . . . or that were people-oriented."[6] Her questions heightened doubt as to whether or how the city could secure the social and economic objectives embedded in its agenda without more resources to address the ways in which racialized disparities in energy burden, housing, and wages are intertwined.

As residents pointed out, clean energy policy alone cannot solve inequality in Atlanta. However, the challenges facing the 100 percent clean energy plan present an important entry point to interrogate how electricity contributed to the racialized uneven development of the city and to reimagine how energy policy and the utility could better serve the city's residents. The final plan released six months later framed the stakes of the city's energy policy in these historic terms: "It is fitting that the symbol for Atlanta is the phoenix. Like the mythical bird that is cyclically reborn from ashes, Atlanta's history is marked by an ability to rise above challenges and emerge stronger and more resilient. From being reduced to ashes during the Civil War to becoming the economic hub of the Southeastern United States, from segregation and Jim Crow to the cradle of the Civil Rights Movement, from a railroad terminus to the most-traveled airport in the world—our progress has been hard-fought but always strongest when built together."[7] The plan, with its stated commitment to working together in a nod to bring Georgia Power to the table, evokes climate change as a "moral duty" and opportunity to move toward a "more equitable, resilient, sustainable, healthy, and profitable" city.[8] While the plan has endured across three Atlanta mayors, progress has been slow going.

Bold claims to make Atlanta one of the most sustainable cities in the US fit within a well-worn pattern of urban boosterism that Atlanta scholars have long observed as a staple of city governance. Charles Rutheiser captures the spectacular process of place-making by Atlanta's urban boosters through both "aggressive, relentless advertising," aesthetics, and design to craft a public image that is always future-oriented.[9] The progressive mythology of Atlanta as a phoenix rising from the ashes of the plantation regime of the "Old South" after the Civil War reverberates in the notion of rebirth, renewal, and growth embedded in New South progressivism advanced by city boosters after Reconstruction.[10] Clarence Stone observes that after World War II and the end of the white primary in Georgia, the hold of the white business elite began to shift with the rising political strength of a Black electorate. Political, business, and civil society actors formed an inter-racial alliance through the urban regime. A commitment to growth and relative racial harmony sutured this coalition that distinguished "The Atlanta Style" or "The Atlanta Way" from other Southern cities roiled by antisegregation protests and the violent suppression of the civil rights movement.[11] Atlanta proudly proclaims its leadership as the cradle of the civil rights movement and the capital of a *New* New South—won and inaugurated by the uprising of Black people. Yet, scholars of Atlanta have also denigrated the urban regime's *New* New South consensus around "business-driven public policy" that deploys public resources to encourage tourism, attract private investment, and grow the metro region, even as growth has systematically disadvantaged poor, BIPOC residents of Atlanta who face gentrification and dispossession as the downsides of improvement.[12]

In this chapter, I investigate the governance of electricity within Atlanta's urban regime. I observe that state capacity to govern electricity emerged in dialectic tension with the utility as the company grew. The utility positioned itself as a quasi-public entity, initially partnered with the city and later pushing outside Atlanta's authority to establish its position as a statewide actor above the fray of urban politics. Georgia Power's predecessors helped to build a legal and regulatory framework that enshrined its authority as a private monopoly regulated at the state level through the state Public Service Commission, eliminating Atlanta's ability to regulate the cost and quality of service. The utility contributed to the segregated urbanization of Atlanta as it worked to grow electricity consumption across the metro region and the wider state, in the process growing its revenues from electric life. As a booster for the region ultimately interested in selling power, Georgia Power deployed a playbook based on low wages, taxes, and energy costs. This paradigm of economic development contributed to the racialized uneven development of the region that energy democracy advocates must contend with today. The utility's entrenched hold over the electric grid backed up by the Commission has constrained creative efforts by city government to decarbonize the city and repair racialized inequality in housing and energy burden linked to the segregated built environment and the sprawling metro region that the utility contributed to building.

The Urban Electric Utility

Atlanta was founded in 1837 on the lands of the Muscogee (Creek) Nation after two wars. At the time of the city's founding, the forced removal of Muscogee and Cherokee people who lived in north Georgia where Atlanta's suburbs would eventually sprawl was on-going. The genocidal dispossession of Indigenous land and lifeways by settlers served the expansion of the plantation economy and the consolidation of local and state government.[13] Initially named Marthasville, the city was formally chartered as Atlanta in 1847, as the terminus for the Western Atlantic Railroad that would open further expansion to the West. It grew haltingly as a peripheral trade and commercial center until after the Civil War. The "rebirth" of Atlanta following the city's partial destruction during the siege by General Sherman has been an enduring touchstone in the city's historical narrative. After the Confederacy's defeat, Atlanta was home to the federal military government and the Freedman's Bureau and was named the state capital in 1867.[14] The city grew quickly with the influx of formerly enslaved people seeking opportunity and escape from new forms of forced labor on plantations. The establishment of universities for African Americans after 1865 also contributed to the growth of the city's Black middle class and position as a Black mecca.

Electricity first emerged in the post-Reconstruction period. W. E. B. Du Bois identified this era as the "counter revolution of property," when Northern capitalists and Southern landowners united to overturn abolition democracy briefly realized through Black workers emancipatory struggle for freedom and the means of reproduction. The refusal to redistribute land and begin a process of reparation to formerly enslaved workers foretold the re-establishment of the white supremacist state. In contrast to the radical possibility of collective ownership over the product of social labor, Du Bois observed that "all of the national treasure of coal, oil, copper, gold and iron had been given away for a song to be made the monopolized basis of private fortunes with perpetual power to tax labor for the right to live and work."[15] Alongside the privatization of land and mineral wealth, new monopolies in railroads, communication, gas, and electricity would exert a hold over working people. Homegrown and Northern businessmen toured the US to recruit investment to Atlanta, presenting the city as a promising investment opportunity and the capital of a New South.[16] Business leaders worked through the chamber of commerce to influence city politicians and "to help the city operate in a business-friendly manner" to realize Atlanta's rebirth.[17] Tax concessions, public contracts, and municipal funding offered to recruit business have remained an enduring strategy for city and state economic development.

Electrification accompanied Atlanta's industrialization. The central business district, Peachtree Street homes of wealthy Atlantans, and cultural institutions were the first to receive power.[18] At the opening night of Atlanta's New Lyceum theater, theatergoers encountered "electric prophets" that were "hanging from the outer walls" and, upon entering the theater, saw a scene akin to the "the proverbial 'blaze of glory'" as under "hundreds of the electric lights the interior decoration was almost dazzling in its brilliancy."[19] The wider public received a taste of this sparkling experience of luxury at fairs that heralded the future of electric life. An *Atlanta Journal* review of the Westinghouse Electric Building at the 1895 Cotton States Exhibition marveled: "Electricity is the power of the world. It has usurped the reign of steam, revolutionized the traffic of land and sea, annihilated distance, and, rivalling the midday sun, it has paled the stars and subdued the shades of night."[20] The seeming mythic power of electricity was domesticated by multiple electric streetcar and light companies that were chartered in the 1880s.

Early utilities vied for lucrative contracts from the city. The first franchise was granted to Brush Electric Company in 1882. The Georgia Electric and Light Company, which would eventually merge with other utilities and streetcar lines across the state to become Georgia Power, followed the year after, receiving permission to operate lighting from 4:30 p.m. to daylight. The first history of Georgia Power espouses the civic-mindedness of the project "motivated primarily by a spirit of pride in Atlanta" and unwillingness to fall behind other cities.[21] Yet, electrification was slow to displace gas, which had been in operation since before the Civil War.[22] Citing the benefits of electric over gas lamps, which "were not lighted when the moon shone, without regard for its stage or the density of the clouds," in 1888, the aldermen awarded Georgia Electric and Light Company a three-year contract for all streetlighting. Electrification imposed a new temporal order on the city center, no longer held back by "flickering and uncertain gas illumination."[23] The constant brightness of electricity opened the frontier of once-dark hours for work and leisure in electric life.[24]

The private provision of electric lighting by companies who sought to cash in on technological progress sat uncomfortably with some in city government. Against the company's protest, in 1897, the Council adopted a resolution to study the merits of building a municipal electric plant. One official noted that "the electric lights should be owned by the city just as much as the water works."[25] However, the review eventually determined the utility's costs and 6 percent profit margin were reasonable, and a new contract for streetlighting was issued in 1899.[26] This time, both Georgia Electric and Light Company and its competitor, the Atlanta Railway and Power Company, bid on the contract, igniting a utility war deemed "the Second Battle of Atlanta."[27]

The fight centered on the two prominent figures within the Atlanta Chamber of Commerce: H. M. Atkinson, president of the Georgia Electric and Light Company, and Joel Hurt, president of the Atlanta Railway and Power Company. Atkinson worked as a cotton sampler and buyer for real estate developer George Washington Adair before becoming involved in banking. He purchased stock in the light company and gained a controlling share with ambition to grow the customer base. In 1892, his company petitioned the city for permission to offer "round-the-clock service" to furnish light and power for building interiors, elevators, sewing machines, saws, motors, and other manufacturing equipment. By 1897, the Company had four hundred customers and supplied part of the power required by the city's electric streetcar lines. Atkinson's rival in the electricity business, Joel Hurt was among one of the many early investors in the streetcars who had overlapping interests in real estate development. As president of the East Atlanta Land Company and founder of Atlanta Trust Company Bank, Hurt set up the first electric streetcar to the eastern suburb Inman Park in 1889.[28] He consolidated and electrified horse-drawn and dummy car routes in what would eventually become the Atlanta Railway and Power Company.[29]

At first the utilities coexisted, with Hurt's Company purchasing electricity from Atkinson's. When the former proposed to build its own power plant and bid on the contract for the city's streetlights, the Georgia Electric and Light Company started a counteroffensive to build duplicative rail infrastructure.[30] The move was purely speculative, admitted Atkinson, writing that: "One has but to look about to find unmistakable evidence of the disastrous nature of the competitions of public service industries. They are natural monopolies. No sane man will put money into such competition. It is financial suicide. It means the destruction of property."[31] Atkinson had been a vocal proponent of regulated monopolies as a member of the National Electric Light Association (NELA) Legislative Policy Committee and conveyed the notion of electricity as a "natural monopoly" to Atlanta. In deploying NELA's talking points to predict the eventual consolidation of the competitors, Atkinson revealed his true intent. His aim was not to compete with Hurt or improve services for the city and its residents, but rather monopolization that eliminated competition altogether and paved the way for guaranteed revenue for the utility through control of the service territory.

An eventual settlement permitted the two rivals to consolidate in 1902, as the newly formed Georgia Railway and Electric Company. As part of its new charter, the utility agreed to pay taxes and recognize continuous transfer for streetcar riders across the city. The merger granted the utility a monopoly over Atlanta and effectively foreclosed competition to secure favorable rates or streetcar fares for its residents. Notably, the final consolidating contract named that if the city ever chose to build its own power plant, tax payments from the company would cease. The threat of lost revenue continued to be used to oppose municipalization, even if consolidation did not quiet calls for public ownership.[32]

The Atlanta Race Massacre and the Birth of a Regional Utility

As Atlanta grew, the question of governing electricity infrastructure shifted from the city to state government. Between 1870 and 1890, Atlanta's population tripled in size and then doubled again by 1910. Demographic changes accompanied the booming population. In 1870, 45 percent of Atlantans were Black. This share would fall to just over a third in 1910, as even greater numbers of white people migrated into the city, and competed for jobs, housing, and municipal services.[33]

The city faced a housing shortage exacerbated by the rigid practices of racial and class segregation. In contrast to the spatial segregation between majority Black neighborhoods in southwest Atlanta and majority white neighborhoods in the northeast of the city that would come to define the Atlanta metro, in the late nineteenth century, there was relative integration of Black and white and rich and poor renters and homeowners segregated by street, block, or individual building. Strict social enforcement of racial separation maintained a physical distance within shared spaces.[34] Yet, the lines between Black and white spaces blurred in the public venues of streetcars, theaters, night clubs, and avenues illuminated with electric lights where racial integration tested the purported tolerance of the New South.[35] The streetcar was unlike anywhere else in the city, explained a self-proclaimed "Northern visitor" in his journalistic study of the color line, because "In almost no other relationship do the races come together, physically, on anything like a common footing. In their homes and in ordinary employment, they meet as master and servant; but in the street cars they touch as free citizens, each paying for the right to ride. The white not in a place of command, the Negro without obligation of servitude."[36] Perhaps given the seeming equality granted to citizens as customers, "the streetcar replaced the railroad as the most visible icon of the aspiring metropolis" and subsequently was among the first sites marred by Jim Crow regulations.[37]

In 1891, the Georgia General Assembly adopted the state's first Jim Crow measure to enforce racial segregation on public transport. The law also gave streetcar operators police powers for enforcement.[38] Black residents organized boycotts when companies tried to institute the law, leading, at first, to limited enforcement by operators interested in maximizing ridership.[39] At the time, the streetcar was a luxury for many workers forced to live in squalid and cramped housing within walking distance to mills and factories where they worked. Populist politicians and newspapers responded to and stoked racist resentment of white workers who could not afford to ride the streetcar while Black middle-class residents could. Preceding the racialized and classed denigration of public transportation that would accompany the rise of car ownership, the streetcar was pilloried as an elitist institution that had fallen into the "hands of grinding monopolies."[40] Racial tension on the streetcars came to a head in the 1906 Atlanta race massacre, later termed "the streetcar riot."[41] It was caused in part by sensationalist newspapers that reported false stories of alleged assaults by Black men on white women.[42] On the night of September 22, 1906, mobs of white people destroyed Black businesses and attacked streetcar riders, murdering an unknown number, estimated between twenty-six and forty-seven Black residents.[43]

The massacre had profound effects on electricity regulation and the racial geographies of Atlanta. Populist politicians and white working-class customers continued to demand public takeover of the streetcars to enforce racial segregation and reduce fares. Instead, at the request of companies, a compromise measure was adopted to regulate electricity under what was then the state Railroad Commission that would "take the street railways entirely out of politics, and render them subject to such regulations as the state may deem proper to place upon them."[44] Georgia became one of the first states to adopt state regulation of electric utilities.[45] Unlike in other states, the measure to expand the authority of what was then the Railroad Commission and renamed the Georgia Public Service Commission in 1922 included a provision for the direct election of commissioners through statewide, popular election. This bill was followed by an amendment to the state constitution in 1908, to disenfranchise African American voters. While regulation proved sufficient to win over the Company's strongest opponents, Cater argues that delegating oversight of the utility to the Commission pandered to "popular demands for the enlargement of white democracy."[46] Utility regulation and white supremacist politics were coupled logics mobilized by the utility to protect and expand their business model. Far from empowering working-class people in Atlanta, this model granted utilities enormous latitude in designing and directing regulation over their captive service territory.

Empowering the statewide Commission to regulate electric utilities cooled but did not eliminate protest, nor did it settle the question of control over utility governance. In 1910, the city council granted a franchise to allow a hydropower company to sell power from north Georgia rivers into the city. While meant to reintroduce competition, the hydropower company and Georgia Power's predecessor soon merged.[47] In 1918, municipalization was raised again when the utility increased rates amid rising inflation and fuel shortages during World War I. A citizen's committee formed to protest the increase as a tax on residents.[48] When the city council tried to prohibit the increase, the state Railroad Commission came down firmly in the Company's favor. The Commission argued that the utility was "the trustee of these great industrial resources, and to this it should receive at the hands of the public and government agencies fair and unprejudiced treatment. . . . The business interests of this section will suffer if they are not adequately served."[49] The Commission's decision to increase rates at the request of the utility and over the opposition of the city has had on-going ramifications that influence Atlanta's efforts to reach 100 percent clean energy today. It cemented the Commission's fiduciary responsibility to ensure the financial well-being of private electric providers that offered a "public service" to citizens of the state and authorized the utility to capture a return on their investment in the private provision of electric life.[50] When the city appealed the Commission's decision, the Supreme Court found that Atlanta and other municipalities do not "have the charter power to pass ordinances upon this subject."[51] The decision hobbled the city's authority to regulate power generated and sold within city limits and formalized state authority over the utility as a statewide entity that was not a part of, but rather acting within the purview of, state government to generate and transmit electricity.

Despite losing any ability to regulate rates, the city tried other tactics to curb Georgia Power's predecessors' growth. In 1919, the Fulton County delegation to the state house introduced a bill to authorize the city "to acquire by purchase, condemnation or otherwise the property of the Georgia Railway and Power Company," and operate its streetcars and gas and electric plants in north Georgia. This bill was strongly opposed by north Georgia towns who feared a takeover by Atlanta more than the private utility. Both the bill for Atlanta's control and a parallel measure sponsored by the Municipal League, which would have authorized municipalities to raise revenue to purchase utility properties, failed.[52] Nonetheless, a decade later amid the Great Depression, the city adopted an ordinance advocating municipal ownership of all public utilities. The potential loss of revenue from the Georgia Power franchise tax was a deterrent to moving forward.[53] Mayor James Key still tried to purchase power for the city from the Tennessee Valley Authority in 1934, but was ultimately unsuccessful.[54]

With its hold over Atlanta firmly cemented, the utility grew across the state. In 1926, the Georgia Railway and Electric Company combined with hydropower companies in north Georgia to form the Georgia Power Company. It would soon extend across half the state's territory, encompassing most urban areas.[55] Over the following decade, the region saw further consolidation with the formation of the Commonwealth & Southern holding company stretching from Mississippi to South Carolina in 1929. While state utilities argued this allowed them to move electrons across the region, the consolidation ran afoul of regulation promulgated under the 1935 Public Utility Holding Company Act. Under order to dispense with gas, transportation, and South Carolina assets from the Securities and Exchange Commission, in 1947, Georgia Power, Alabama Power, Gulf Power, and Mississippi Power were reincorporated under the Southern Company.[56]

The question of who Atlanta's electricity provider would be was effectively laid to rest. In 1965, seemingly without debate, the city signed a ninety-nine-year franchise agreement to purchase energy from Georgia Power. The 1973 Territorial Electric Service Act fixed the boundaries of electric utilities across the state.[57] State regulation locked Atlanta into a position of deference to the utility as part of the service territory from which it was guaranteed customers, even as the utility would fail to serve Black Atlantans in particular.

Streetcars and Segregation

Just as the Atlanta race massacre set in motion a series of legislative acts that changed the structure of utility regulation, so too did the massacre shape the negotiation over public space, housing, and public services in Atlanta, including electricity and electric streetcars. In its aftermath, the city's white elite gestured to work with Black leaders to restore Atlanta's image. Conservative reform aimed to quash a more militant response from Black people who armed during and after the massacre to defend their homes and institutions against white aggressors.[58] Measures negotiated between Black and white elites to prevent further violence expanded race-based zoning. The city council adopted racial segregation ordinances in 1913 and 1917 although both were deemed unconstitutional by the Georgia Supreme Court.[59] Designated areas for Black and white residents in the 1922 zoning plan were also voided. Nonetheless, legal tools including zoning to establish commercial, industrial, or highway buffers between Black and white communities and racial covenants as well as extralegal terrorist violence, formed what Akira Drake Rodriguez terms "white supremacist spatial logics" that organized the production of urban space to advance the comfort and wealth-accumulation of white Atlantans.[60] Electrification mapped onto and wired the existing spaces of the city such that the utility contributed to demarcating sociospatial boundaries as it produced new urban spaces and moments of programmed consumption within everyday life.

In an effort to sanitize the city after the massacre, city officials used electric lighting to symbolize "progress and cultural advancement" and as a tool of policing and surveillance to deter criminal activity in new public spaces.[61] While historian David Nye argues that electricity democratized urban spaces and gave shape to a mass society, Tera W. Hunter corrects that in the New South "democracy in the parks was reserved for whites only." Amusement parks and the "Great White Way"—an electrified streetscape constructed at Five Points in Atlanta to draw patrons back into public and commercial spaces—enforced Jim Crow and employed electricity to illuminate wholesome, "white consumer fairylands" distanced from the perceived threat of Black people.[62] Black residents frequented their own "White Way" on Auburn Avenue. Electricity also lighted Luna Park and dance halls and saloons on Decatur Street that were "a kaleidoscope of light, noise and bustle from dawn to dawn," offering escape and pleasure outside the moralizing and rigid structure of working days.[63]

Beyond lighting the segregated spaces of leisure, the electric streetcar reinforced white supremacist spatial logics in Atlanta's growing residential footprint. The advent of streetcars made suburban developments accessible to the downtown business district and spurred new residential construction.[64] The Company history explains that "the street railway made possible the City's growth in area. The extension of lines into suburban territory added to the health, comfort, and happiness of the citizens by making it possible for them to work in the City and reside in the suburbs where land and living costs were cheaper. At the same time, the street railway lines greatly enhanced the value of suburban areas."[65] Proximity to streetcar lines and electric hookups, water lines, and other "modern conveniences" were used to sell lots and homes.[66]

Electrification intersected with racialized labor and property markets as wealthier, whiter communities were first to receive electric power and, hence, benefit from the increase in property values that came from having a wired home and access to transit.[67] The first ring of suburban development in planned park-neighborhoods catered to those seeking to escape the crowded city and soon established an aspirational norm for housing patterns even as park-neighborhood building practices bared white working-class residents who could not afford large homes and minimum lots sizes while racial covenants prohibited sales to middle-class people of color.[68] Black property developers built in the south and west of the downtown with "unofficial agreements" from the white elite.[69] Cooperation from white municipal leaders was necessary to negotiate for other infrastructure such as roadways, streetlighting, parks, and transit. Segregated neighborhoods that were "products of the twin forces of exclusion and seclusion" allowed Black people to make space "to bolster their autonomy and collective power and to escape exploitation by whites."[70] Black women would also organize through neighborhood associations and tenant unions. Their work had a profound influence on city planning and raised attention to the lived experiences of urban infrastructure, later including some of the first measures to recognize and address energy burden.[71]

As the city grew, the quality of transportation became an increasing concern. Since the 1891 Jim Crow ordinance segregating streetcars, Black Atlantans had protested second-class service. The Colored Jitney Bus Association started in 1922 and operated bus lines along streetcar routes to alleviate crowding and provide more comfortable service for Black patrons. It was put out of business after the city council prohibited buses with fewer than seventeen seats from operating at the urging of the utility.[72] Two years later, when Georgia Power sought to increase their rates for streetcar service, Black customers petitioned the Commission to deny the increase because "Georgia Power was not giving good service to all the citizens of Atlanta." Arkwright, who personally litigated the case, sought to stop the testimony of the fifty men and women who had come to the hearing, arguing that the streetcars empowered people with the ability to pay who "have a chance on the streetcars to make a display of their equal rights more so than they do elsewhere." In contrast to the Georgia Power president's defense of "separate but equal," a Black doctor and a minister who were permitted to testify shared that they "had been roughly and unjustly treated by street car employees."[73] In seeking to silence Black customers, who were at that time already largely disenfranchised, Georgia Power aimed to fast-track a rate increase that would extract revenue from residents dependent on the private monopoly for transport to and from work, which unions, public officials, and white civic leaders also opposed. A partial increase was eventually awarded.[74]

Inadequate transportation services continued even after Georgia Power and other private transportation companies introduced buses and trackless trolleys. Many who looked further outside the city center to realize the dreams of homeownership had to rely on multiple transfers to get to work and new Black subdivisions had to fight for bus service from Georgia Power and other private transit lines.[75] One resident wrote the mayor, imploring that Black residents moving in search of better housing should not be penalized with excessive transport fares: "Many steady, hard-working colored couples have long wanted to buy lit-new clean houses instead of living at a rental of over fifty dollars monthly in half of a dilapidated house discarded by white people."[76] Price gouging was one of many injustices of segregated transportation. Alice Adams, a domestic worker employed in Druid Hills described the time she spent commuting as exhausting and emotionally draining. The streetcar added additional labor to her long days of work as a housekeeper.[77] Black patrons reported routine policing, including use of force by white streetcar conductors to enforce ambiguous, yet segregated seating arrangements. In separate incidents, streetcar drivers murdered Black veterans Madison Harris and Walter Johnson and wounded John R. Owens in 1946. All three cases were dismissed before a jury trial by the same judge.[78]

Black transit riders were in revolt by the 1940s. The utility corresponded with ministers, the NAACP, and Black and white riders who protested the violent treatment of Black passengers.[79] In 1942, the Georgia Interracial Commission provided recommendations to Georgia Power to improve race relations. The utility and other transportation companies committed to reassign "problematic" drivers from routes used mostly by Black customers.[80] Georgia Power sold their stake in the streetcar system as required by federal regulation in 1947,[81] just before church leaders led a campaign to desegregate transit in 1955. Mayor Hartsfield supported the effort, hoping to avoid the bad press from the violent repression of civil disobedience seen in other cities across the South. Fearing this could hinder commerce and convention trade in the city, he famously asserted that Atlanta was "a city too busy to hate."[82]

Mayor Hartsfield's efforts to build interracial cooperation between Black civic and faith leaders and white businesspeople became the cornerstone of the "Atlanta Way" that has since characterized the urban regime of the *New* New South. The Atlanta Way propelled the city and state forward for national business investment. Georgia was among the "states with reputations for slightly more moderation" on racial integration and benefitted from significant industrial investment and gains in manufacturing employment in the 1960s.[83] The largely pragmatic business decisions of the elite did not repair inequity built into urban infrastructure that shaped the quotidian experience of electric life for Black Atlantans.[84]

The pattern of suburbanization that electric streetcars first enabled accelerated as working-class and middle-class white people resisted the mayor's efforts to desegregate neighborhoods, schools, parks, and buses through physical violence and distancing, leaving the city for newly constructed suburbs while businesses and tax dollars followed.[85] Physical relocation was a propertied investment in whiteness powered by mushrooming energy consumption. Private vehicles—that may in the near future be electrified to support greenhouse gas emission reductions—carry commuters to privatized residential spaces where enormous amounts of electricity are required across the "sprawling sites of social reproduction."[86] Environmental histories of Atlanta concentrate on the central importance of automobiles and road infrastructure in establishing racial boundaries between the suburbs and the deindustrializing urban core.[87] This pattern of suburbanization was first made possible and expanded through the streetcar. As car ownership increased, streetcar lines were paved over and forgotten. Public transit, which was once the envy of white, working-class Atlantans, was denigrated. Since Georgia Power's transit system was turned over to the Metro Atlanta Regional Transit Authority, suburban counties adjacent to Atlanta have mobilized racist and classed rhetoric about crime to fight attempts to expand public transportation. White supremacy continues to impede public transit as a collective project of social reproduction in Atlanta.

Atlanta's population began to fall in the 1960s and did not rebound until the 1990s as more affluent residents returned, sparking gentrification. The city surpassed its 1950 population in 2020, reaching 496,000 total residents. In turn, the population of the metro area increased from just over 660,000 residents in 1950 to more than 6.1 million residents today. Into the 2020s, Metro Atlanta's population grew faster than anywhere else in the US, except the fellow Sunbelt cities of Dallas, Phoenix, and Houston.[88] It also remains among the most segregated cities in the US and consistently ranks highest in income inequality.[89] Even as Georgia Power lost control of Atlanta's public transit, as manager of electricity infrastructure, the utility would continue to profit from its monopoly hold over Atlanta as the metro sprawled outward.

Selling Georgia

Electricity powered metro Atlanta's growth with Georgia Power proudly campaigning across the US to recruit residents and industry to "Move Forward, Move South."[90] This aspirational promotion of an electric way of life, made comfortable by refrigeration and air conditioning that could control the seasons, wedded modernization of homes and businesses to electricity consumption. In 1927, Georgia Power established a Department of Community and Industrial Development, including a recruitment office in New York City where information on industrial sites could be easily shared with "eastern and northern plant executives."[91] This promotional material was also packaged in the company's first annual report in which the utility rewrote the history of the plantation "Old South" as an idyllic, rural landscape until "the war between the states left it a desolate, smouldering [sic] waste." Carrying forward the idea of a New South as a transformed, if not repentant, geography based on a racist and xenophobic settler imaginary, Georgia was presented as an untapped market with bountiful resources where investors could escape the cold and labor militancy of European immigrants:

> Comparable with the development of the West in the latter half of the last century is that which the South is experiencing now. . . . Today a vast region lies waiting the pioneer and prospector. . . . Many factors have had a part in the attraction of the Nation's attention to the Southeast, and especially to Georgia. Outstanding among these is a temperate climate that offers healthful living conditions and manifold advantages to business and industry. Other factors include decided advantages for manufacturing: a fine class of Anglo-Saxon skilled and unskilled labor, giving freedom from labor troubles; proximity of raw material; low taxes; abundant power resources; almost virginal agricultural lands; splendid educational institutions; sound banking facilities; inexhaustible natural resources; adequate means of transportation and shipping, and a cooperative spirit throughout the State.[92]

The utility pointed to their own segregated workforce as example of the "fine class" of workers employers could find for hire (see figure 1.1).

The increasingly aggressive promotion of a Southern style of doing business and a low-wage and nonunion workforce follows a trend James C. Cobb describes as a "crusade for industrial development" that pioneered development bonds, tax breaks, and subsidized worker training and infrastructure to land mobile capital.[93] Georgia Power's Better Home Town competitions encouraged planning for infrastructure and industry.[94] The utility in turn gained information to advertise what Georgia had to offer through national newspapers and brochures targeting specific industries.[95] This collaborative planning with local governments also pushed Georgia Power into new rural communities that the utility had not previously connected to the grid, eventually reaching 155 out of 159 Georgia counties.

Figure 1.1
Georgia Power's workforce in the 1927 company yearbook. Photo Credit: Kenan Research Center at the Atlanta History Center.

A comprehensive study of new manufacturing plants in the South produced by the National Planning Association in 1949 reported that natural resources, access to markets in growing Southern cities—as well as their ports and rail lines—and labor availability and attitude were the reasons cited for plant location. The age, gender, race, and skill level of workers were all considered. Some textile operators reported a preference for rural communities where they would be the only employer, making unionization less likely. One explained: "We're not running away from unions, we're just staying away from them."[96] With few exceptions, "electric power alone attracts few industries to a region," but the "general locational importance of electric power is best described as 'improving the industrial environment.'"[97] A subsequent survey of town mayors and chamber of commerce officials reported that Georgia Power and the state Department of Industry and Trade were the most important economic development agencies in Georgia. And a corresponding 1972 survey of companies concerning seventy-one plant location decisions showed that "local organizations had been more involved in the forty-eight midwestern plan location decisions, but state, utility and railroad promoters had been more active in the South."[98] Utilities intended to make real on a slogan used by Southern Company in a national ad campaign that "the last half of the twentieth century belongs to the South."[99]

Cheap electricity became part of the conjuncture pushing forward southern industrial policy and urbanization.[100] Some of the first heavy industries to locate in Atlanta were branch plants, like the Ford Motor Company factory in Hapeville. The region also captured federal investment that flowed through military contractors, army bases, and shipyards, which was directed South with the intent to promote development and reverse persistent poverty.[101] Electrified firms reversed the centralization of urban production that characterized the advent of coal-power in the nineteenth-century Industrial Revolution.[102] Hamer describes that electric power

> eliminated the cumbersome arrangement of powered machinery around a central source of steam power by allowing small motors to be positioned quite independently of the power source. . . . That arrangement, especially where mass production was concerned, no longer tolerated crowded, multi-storied settings. It called instead for land-extensive settings. . . . In the past only rather self-sufficient, massive works like steel mills, often in search of nonurban raw materials, populated what today are the suburbs. Now the industrial exodus was on.[103]

The demand for space to efficiently lay out electrified plants unfolded alongside growing demand for residential housing in the suburbs and worked to push the borders of the Atlanta metro region further outward—at the same time as Black Atlantans demanded a greater share of the social wage. New factories located out from the urban core of the city, accelerating the sprawl that stretches across the twenty-nine-county metro area of Atlanta today.[104] The share of workers employed in manufacturing in Georgia peaked at 27.4 percent in 1955, falling to 6.6 percent in 2022 as employers automated or moved to even lower-wage locations.[105] New job creation clustered in finance, insurance, real estate, communications, government, services, and retail was highly bifurcated between high and poverty-level wages, and service industries in the city also faced competition from retail and hotels located in the suburbs.[106] As real estate investment, corporate headquarters, and operations suburbanized, investors turned away from Atlanta to "spend more time engaging with a different set of political leaders" in state and suburban county governments.[107]

Georgia Power, which is both deeply invested in and connected to every other industry dependent on its product, had a shared leadership role within the urban regime and the wider regional bloc that guided the South's transition from agriculture to industrialization.[108] Georgia Power's executives sit on commercial, philanthropic, and public boards for the city and wider metro area, including Central Atlanta Progress, which promotes downtown Atlanta; the Atlanta Regional Commission, which is the planning agency for an eleven-country area; and the Metro Atlanta Chamber of Commerce, representing a broader swath of businesses across twenty-nine counties. In a 2004 interview reflecting on the chamber's growing influence in regional planning, a Georgia Power official described: "We do have significant involvement. . . . It's a benefit to us. We want people to know we're not just the electric service provider."[109] Beyond providing power, the utility lobbies to maintain the organization of state capacities that keep Georgia a top state for business from tax credits to lax environmental policy, noting that growing any kind of industry grows electricity demand and hence revenue to maintain the wider grid.[110] The utility also brokers public-private business deals. Its Community and Economic Development arm conducts research for city and metro agencies and provides complimentary maps, renderings of proposed sites, market research, and information on available tax credits and incentives for potential investors.[111]

Even as the Atlanta region has grown in national and international influence, Georgia Power maintains a strong influence over local and state government. Hugh, an energy consultant, explained in an interview that "utilities across the country are political powerhouses, there's no question about that. . . . But I'd argue in the South, they're even more politically adept than in other parts of the country."[112] Georgia Power works through the city, metro-regional coalitions, and at the state and regional levels, asserting itself to be a public utility working in the public interest. Critics have rightfully asked who Atlanta's unbalanced economic development serves. The crusade for development is a race to the bottom in inter-regional competition that reallocates public resources to industry, eroding the corporate tax base and transferring taxpayer and ratepayer dollars to build infrastructure for corporations, all while stripping funding from broader public infrastructures.[113] Electric life that the utility manufactured to grow Atlanta remains tethered to regional and racial inequality.

Urban Austerity, Urban Sustainability

The growth of the Atlanta metro region, spurred by cheap land, labor, and energy, failed to achieve the broadly shared economic prosperity envisioned in the *New* New South. Against the sprawling backdrop of the metro suburbs, in the 1990s, Atlanta began a series of climate and clean energy programs that inaugurated a new boosterism around urban environmentalism. Cities across the world adopted aggressive sustainability programs to do their part to address international climate concerns. Climate urbanism or urban environmentalism was trendy and, identified Peter Brand, emerged out of very real material constraints in postindustrial cities facing decades of declining population, tax revenue, and federal funding that hobbled efforts to take on other structural reforms to improve reproductive infrastructures of housing, healthcare, or schools. Urban environments became "an arena in which social welfare could, to some degree or another, be discursively constructed and materially produced."[114] In Atlanta, environmental programs executed through public-private partnerships became a cornerstone of economic revitalization.

Atlanta's successful bid to host the 1996 Olympics was an early moment of climate change awareness. The Turner Foundation, under the direction of CNN founder and media mogul Ted Turner (whose grandson currently serves as Atlanta's Deputy Chief Sustainability Officer) and his then wife, Jane Fonda, encouraged the city to invest in solar manufacturing to "sell panels nationwide, provide jobs to [Empowerment] zone residents and equip publicly owned facilities such as the new Olympic stadium with an environment-friendly energy source."[115] Debuting an innovative approach to infrastructure on the international stage could make Atlanta the "Solar City."[116] The proposal to pair reindustrialization and employment in solar panel manufacturing targeted to the most impoverished neighborhoods of the city failed to materialize. While Georgia Power sponsored electric buses at the Olympics, little else emerged from this early vision, and subsequent financial support from the Turner Foundation for municipal energy planning was rejected by the city in favor of assistance promised by Georgia Power.[117] At the same time, "urban revitalization" measures that were implemented for the Olympics included demolition of public housing and antihomeless policies that evicted poor and working-class people from the city center, setting in motion gentrification.[118]

Urban environmentalism became ascendant in city planning under Mayor Shirley Franklin. She signed onto the US Mayor's Climate Protection Agreement to honor the carbon reduction targets set in the Kyoto Protocol, began construction on water and sewage infrastructure to bring the city into compliance with federal water quality standards, and initiated construction of the Beltline, a series of parks integrated with walking and biking trails that circumnavigate the city center.[119] Following redevelopment strategies initiated around the Olympics, the Beltline was a major public-private partnership "built on the back of expected rising property values, to reshape the city, both physically and demographically."[120] Intending to connect white and Black and poor and wealthy neighborhoods through twenty-two miles of trails and parks, the Beltline also fueled speculative investment that drove up property values without sufficient measures to protect residents from rising rents, insurance, and property taxes. Community organizations, including Georgia STAND-UP, mobilized residents to engage in the planning process, negotiate a community benefits agreement for the Beltline, and engage the newly formed Office of Sustainability.

In 2009, Mayor Franklin appointed the city's first sustainability director. At that point, explained Rebecca, a former city employee, it was "the first round of big city mayor's doing sustainability. . . . New York City, Chicago, Seattle, San Francisco . . . Boston [were the] major cities with sustainability plans, and staff," and Atlanta wanted to be part of it. As a newly established political division, the office was not guaranteed to last past the Franklin administration, and efforts focused on building relationships with staff and internal processes, like "how to help City Hall just be turned off at night."[121] The division partnered with university students to benchmark energy use and show where the city could save money. Brad, a former city employee, explained that "the arguments that were made to leadership, whether it's council or the mayor fell under a convenient little saying that was 'green had to be green.'"[122] The American Recovery and Reinvestment Act (ARRA), in 2009, provided resources for implementation that prioritized "green job" creation. Atlanta seized on this brief reversal of the prior decades of federal austerity to chart expansive plans encompassing everything from transportation to air quality. Without ARRA funds, Rebecca remembered, "we would never have been able to get the money from the city because we were initially funded by pure philanthropic dollars. . . . [and] we didn't have enough money to do projects. So the ARRA money came at the right time, to be able to put cash behind our dreams."[123] Funding for initiatives remained a perennial concern, yet the office grew to more than twenty employees under Mayor Kasim Reed, who published the first Sustainability Plan in 2014, *Power to Change*, and hired Stephanie Stuckey, a former house representative to lead the office.[124]

Atlanta endorsed increasingly ambitious clean energy targets. In 2017, a city councilmember and mayoral candidate, Kwanza Hall, dismissed climate change at a candidate's forum. To make up for the public gaffe and show he understood that climate change was both real, and a serious issue for the city, he proposed a 100 percent clean energy commitment. The resolution, which applies to all electricity used within city limits, including municipal operations at the city-owned Hartsfield–Jackson Airport, was unanimously adopted by City Council.[125]

The ability of the city to reach this ambitious target has been met with skepticism, not least from city staff who were not consulted on the resolution before it was approved.[126] Tasked with figuring out implementation after the fact, they pulled together a team from national and local nonprofits to develop and workshop what became three decarbonization scenarios: business as usual, achieving 50 percent local clean energy, and maximizing Atlanta's local clean energy potential to implement energy efficiency and rooftop solar. Despite attempting to assert its own energy future, the plan conceded that any of the three options would require the city to purchase renewable energy credits because Georgia Power has been unwilling to bend to the city's resolution and is far short of the Southern Company target to decarbonize by 2050.[127] While the city signed a Solar Energy Procurement Agreement in 2017 to install solar on city-owned properties, reaching its broader target for all Atlanta residents remains at the utility's discretion.

Given the restrictions Georgia Power's monopoly on energy procurement presents, the city has consistently prioritized energy efficiency in its sustainability programming. Across Georgia, the National Renewable Energy Lab estimates 26 percent of current energy use could be saved through cost-effective efficiency improvements.[128] In 2011, Atlanta signed onto the Better Buildings Challenge, a voluntary initiative among downtown businesses to reduce energy and water consumption in commercial buildings by 20 percent below the 2009 baseline. Relying once again on federal pandemic-era funding through the American Rescue Plan, in 2023 the city piloted a $2.4 million WeatheRISE Atlanta program to provide whole home retrofits to income-qualified households.[129] Such programs by the city work around the utility, which has made only marginal investments in energy savings. In 2021, Georgia Power invested approximately a third of the national average across all large utilities in energy savings.[130] Limited funding is spread across the utility's entire service territory, meaning very little is available to Atlanta specifically. Brad described that the utility's opposition to energy efficiency "make[s] everyone's bills higher. . . . And it amounts to a taking, from the customer to the company. . . . We're leaving billions on the table every year that should be staying in our citizens pockets, staying in the state of Georgia's economy, being put to higher, more productive uses, and reducing our overall public health costs."[131] In 2022, Georgia Power reversed in its historic refusal to see energy efficiency as a resource and included it in modeling alongside other energy resources in its long-term procurement planning through the Integrated Resource Plan (IRP). Nonetheless, continued underinvestment remains a "taking" that appropriates customer's money and time in wasted energy.

The city has limited recourse to challenge the utility's resource decisions. In 2019, Mayor Keisha Lance Bottoms's administration did not intervene in the IRP in which the utility's resource decisions are evaluated by the Commission, fearing retaliation by the General Assembly, which had threatened to take over the Atlanta airport that year.[132] The threat of preemption chills city action and hobbles programs that support working-class people's needs. Since the state preempted city regulation of the utility a century prior, the state legislature has preempted city powers to increase the minimum wage, mandate paid leave, require fair scheduling, encourage collective bargaining, implement rent control, provide sanctuary to immigrants, and regulate residential gas hookups.[133] Despite these real constraints and perceived threats, when Mayor Bottoms deprioritized environmental programs inherited from her predecessors, many staff members resigned. Linda, a sustainability officer in the city, described the attrition as an enormous loss. Under the Reed administration, "This city felt to me exciting and vibrant and moving in a progressive way in a way that we could have risen to the top of this country. I mean, it doesn't feel that way now."[134] She believed the boosterish rhetoric of Atlanta being a "world-class, sustainable city" to be possible with the right leadership in place throughout government that could act on ambitious policy and challenge the hegemony of the utility. In 2022, newly elected Mayor Andre Dickens appointed Chandra Farley—a grassroots energy justice leader—as chief sustainability officer.[135] Farley's leadership positioned Atlanta to take advantage of IRA-funded Solar for All and energy efficiency programs. While Mayor Dickens revamped city support for issues of climate action, he also worked to repair relationships with state officials and conceded to other priorities of the urban regime, including the environmentally destructive project to bulldoze forestland and build the Atlanta Public Safety Training Center, derisively known as "Cop City."[136]

Atlanta is not alone in its 100 percent clean energy commitment and has begun to work in concert with Decatur, Savannah, Athens–Clarke County, and Dekalb County as part of the Georgia Coalition of Local Governments to push for clean energy at the Public Service Commission. This coalition is indicative of the shifting position of Atlanta within the state. Larry Keating argues that while the "capacity of the business elite to shape public policies and subsidies is still the dominant force in Atlanta politics, Atlanta is no longer the sole focus of local political activity."[137] The growth of the surrounding suburbs, which contain 90 percent of the metro region's population, had led to the diminishing importance of Atlanta's city government in what has become a regional regime. The city remains the cultural and political capital but must work in coordination with metro and state political and business leaders to influence Georgia Power and energy procurement for the metro region and the state political bloc.

In 2025, Atlanta was 18 percent of the way to achieving its 2035 goal for 100 percent clean energy,[138] yet, decarbonization is only a piece of democratizing the energy system. To address racialized, uneven development that electricity capital helped wire into urban life requires reparative investments to address energy burden. Control over energy planning by the utility and Commission means that the equity mandate of Atlanta's clean energy ambitions are ignored or enacted in piecemeal fashion. At the same time, gentrification—spurred in part by residents seeking "green" amenities such as public transit, walkability, and parks like the Beltline—means that former residents of the city who have been pushed into the increasingly racially and economically diverse suburbs are no longer able to benefit from limited clean energy programming the city has been able to fund.[139]

Climate and environmental ambitions remain fettered to the public-private negotiation of the urban regime and the regional bloc that influences state politics. This system of elite management in which democratic demands move slowly, if at all, through the relationships of public-private negotiation within and outside the city affects the tenor and possibility of local political action. Rodriguez finds that over time Atlanta's system of "biracial cooperation" has "create[d] political opportunities for the biracial elites in power," while gutting venues for democratic participation for working-class people to win substantive improvements in their lives.[140]

Voting Rights and Representation

In 2020, the stakes of having a utility concerned about the welfare of Black and working-class people increased when the COVID-19 pandemic began. In June 2020, the Commission allowed Georgia Power's voluntary moratorium on shutoffs to expire, threatening disconnections for households at a time when one hundred thousand Georgians, many of whom had been laid off or lost work due to the pandemic, were behind on their bills.[141] Of all utilities required to report disconnections nationwide, Georgia Power would shut off more people's power than all but one other utility between 2020 and October 2022. Notably, Georgia Power's arrears—the customers' debt that could have been canceled to prevent shutoffs—totaled less than 1 percent of the amount paid to shareholders in dividends during this time.[142] Racial disparities in COVID-19 infection and death as well as job loss were well established, yet, disregarded by the Commission, even as Atlantans marched in the streets for Black Lives following the police murders of George Floyd and Rayshard Brooks.[143]

In response to the Commission's disregard for the unaffordability of electric life in Atlanta, on July 14, 2020, four representatives from the NAACP, Black Voters Matter, and the Georgia League of Conservation Voters filed a lawsuit alleging that the statewide election of Commissioners leads to the underrepresentation of Black voters in violation of the Voting Rights Act. The lawsuit cited the same history elaborated here, which traced Georgia's system of electing Public Service Commissioners to white supremacist, populist fervor in the early twentieth century.[144] It further named that while Commissioners are required to live in and represent specific districts, at-large voting means that the statewide white majority can "defeat the candidates preferred by black voters."[145] Black people make up more than 30 percent of Georgia's population, yet at the time, all the Commissioners were white and Republican. Since the Commission was first established in 1879, only two African Americans have served on the Commission. Both David Burgess (appointed in 1999) and Fitz Simmons (appointed in 2021) were first named to the Commission by the governor to fill a vacancy and not initially elected to office, following a pattern of "informal slating" that confers incumbency on a hand-picked candidate before a competitive election.[146]

In August 2022, a federal judge sided with the plaintiffs and ruled that the election method dilutes the votes of Black people, which halted the 2022 elections, until a new voting method could be determined.[147] The decision was overturned on appeal and the US Supreme Court declined to take up the case, meaning that statewide election for Commissioners will continue. Since 2022, the Commission has continued with the same members after elections in 2022 and 2024 were stayed. The General Assembly adopted a bill to extend the terms of the Commissioners until two elections could be held in 2025.[148] While unsuccessful, these efforts to change the method of electing commissioners endeavored to disrupt the more-than-century-old system of regulation that enriches the utility and advances its growth at the expense of Atlanta and cost-burdened residents whose time, energy, and budgets are taxed.

Conclusion

Despite the stranglehold that corporate and state governance has over the city's electric grid, environmental and community organizers have found common cause in asking how energy democracy could be practiced in the city. Demands for clean and affordable utilities and improved public service connect to urgent concerns about the crisis of climate change and are carried forward by a cross-class and inter-racial movement insistent that elected representatives act to mitigate the environmental effects and financial burden of electric life. This organizing takes aim at multiple scales of the utility's influence from households, industry, urban infrastructure, and the Commission through which electric life and the racialized uneven development of Atlanta are defined and reproduced. Pushing back against the Atlanta Way that has organized the urban regime, state power, and social reproduction to support growth, new political imaginaries emerge through city sustainability planning and energy democracy organizing to influence it. These efforts begin from the everyday uses of electricity and implicitly embrace a feminist urban political ecology in asking how energy and public institutions can support social reproduction.

2 Paying for Electric Life

In the crowded hallways of the Reagan World Trade Center building in Washington, DC, clean energy industry executives shared coffee and mingled with US Department of Energy (DOE) officials at the inaugural Demonstrate, Deploy, Decarbonize conference hosted by the agency's Loan Program Office (LPO). The conference showcased federal financing and research aimed to "lift off" decarbonization across the economy and urge innovation in hard-to-abate industrial sectors. Equipped with $400 billion in lending authority from the Bipartisan Infrastructure Law and Inflation Reduction Act, the conference demonstrated how LPO loans could work alongside grants and tax credits available from the federal government to push forward an energy transition that is "private sector-led and government-enabled."[1] First established in 2005 under the Energy Policy Act that authorized low-cost loans for novel technologies that "avoid, reduce, or sequester air pollutants or anthropogenic emissions of greenhouse gases," LPO has seeded projects from Tesla to Southern Company's Plant Vogtle expansion. Just a month before the conference, Unit 3 of Vogtle came online.[2]

The conference was abuzz with the news. Standing on stage in a TED Talk–style presentation before a massive auditorium, an LPO advisor described his visit to see the facility as awe-inspiring to see "just how big this thing is. . . . [where] turbines the size of sport utility vehicles" generate power. He heralded Vogtle as a triumph of public-private collaboration and a cautionary tale in moving forward with advanced nuclear energy as it "took too long, [and] cost too much to build."[3]

The following day, in a much smaller meeting room, conference attendees gathered to dig into the details of the DOE's liftoff report for advanced nuclear generation. The report highlighted pathways to finance 200 gigawatts (GW) of new nuclear energy, which the DOE saw as a necessary baseload technology to decarbonize the grid and meet the US's Paris climate commitments.[4] The discussion centered on how LPO could support the future of the nuclear industry through a special-purpose vehicle to finance several utilities working together to build small modular reactors. This would enable companies to share the efficiencies of scale learned in building, testing, and repeating one design to mitigate cost overruns that have plagued nuclear construction since its inception. Without addressing what an increase in costs spurred by investment in the rebirth of a nuclear industry would mean for the ordinary customers who already struggle to afford their monthly bills, another LPO senior advisor and coauthor of the liftoff report described that "this might require a North Carolina ratepayer shouldering some of the cost for a Virginia customer."[5] Notably, the hypothetical proposal she offered obligated customers in the South where Southern Company, Duke, Dominion, NextEra, and Entergy own and operate monopoly, vertically integrated, regulated utilities. As she transparently described: "We're not going to get to the scale we need without the traditional regulated utilities." Southern utilities, operating under what Conor Harrison terms a strategy of "monopoly accumulation," are among the most profitable in the country and have benefited both from selling power to customers in their own captive territories as well as from operating generation and transmission assets in other states where the vertical integration of electricity service was unbundled following electricity deregulation in the 1990s.[6]

In deregulated states where electricity rates are priced through markets, new nuclear is not cost competitive. For regulated utilities, rates are set by state public service commissions based on rate-of-return regulation. This means that utilities build capital assets to generate and distribute power, and the regulator determines rates based on both the value of the utility's capital investments to produce electricity and an allowable rate of return on their fixed costs—or percentage profit—to be distributed between the company and shareholders as dividends. Within this system, monopolies are incentivized to overbuild on "gold-plated" projects like nuclear that allow the utility to capture maximum rents and, thus, maximize shareholder value.[7] The guarantee provided by regulators in the noncompetitive Georgia Power service territory means that Vogtle is one of LPO's "highest rated projects." The LPO advisor continued, "Because they're a regulated utility their repayment is guaranteed," downplaying the potential risks of nuclear construction and cost overruns to nervous companies in the room. In effect, she described, for the regulated utility, future demand for electricity can be collateralized as a financial asset. The company tabulates its projected electricity load, estimated from aggregated household, commercial, and industrial demand, to calculate how much capacity it must build. These daily uses of power for electric life in a captive service territory guarantee future repayment to the American taxpayer who funds LPO, as well as stockholders and bondholders who lend the company money to finance new infrastructure. In other words, regulation allows Georgia Power to wager the life-times of customers whose future, anticipated electricity demand is offered "as *risk-absorbing collateral*" to LPO, banks, and other lenders.[8] Customers in Georgia Power's captive service territory are then obligated by the Public Service Commission to pay off the debts that Georgia Power owes on their monthly bills.[9] While risk is collectivized, the mechanism that allows monopoly utilities to exact a generous rate of return transfers costs to ordinary ratepayers who must work additional hours, make changes to their home energy consumption, or forego necessary expenses when they face an increase in the light bill.[10]

In Georgia, the Commission describes itself as having a greater "impact on people's lives" than most other government agencies given its direct influence over electricity and other regulated utilities including gas and telecommunications.[11] Its oversight of electricity is premised on the social necessity of power in everyday life that is too important to be left to an unreliable provider. The monopoly utility is required to provision service to all customers within its service territory. In turn, the Commission is empowered with a narrow mission as a technical regulator to "ensure that consumers receive safe, reliable and reasonably priced telecommunications, electric and natural gas services from financially viable and technically competent companies."[12] This mandate thus ensures the financial viability of Georgia Power, yet excludes social and environmental concerns of racial disparity in energy burden or pollution exposure, worker's rights, or climate change.

The question that this chapter asks is not whether Plant Vogtle should have been built. Rather, following the argument that the investor-owned, regulated utility is the only financial mechanism to build novel, risky, or big electrical infrastructure at the scale needed to address the climate crisis, I consider how this system has failed to ensure affordability for all and review how working-class, disabled, and elderly people have advocated for change at the Commission to improve social protection. I expand on Richard

Hirsh's analysis of the utility consensus between regulated utilities and their regulators that incentivized growing and building, to look specifically at rate-of-return regulation. Regulation encourages growth and makes the utility profitable, even at the expense of customers who struggle to afford the conveniences of electric life. In the first part of the chapter, I review utility advertising programs to increase everyday energy consumption. I then turn to consumer advocacy campaigns, beginning in the energy crisis in the 1970s that preceded the contemporary movement for energy democracy yet utilized similar tactics to mobilize poor and working people to speak out about the everyday impacts of the utility's growth mandate. This chapter traces three campaigns that strived to lower the cost of electricity, stop the addition of fees for construction work in progress, and reduce base fees charged to all customers. This history begs the question as to how the Commission could be empowered to regulate Georgia Power, how power could be priced in a progressive way that supports poor and working people, and whether a genuinely public utility could better serve Atlanta residents, an idea I return to in the conclusion.

Throughout, I evaluate this process of rate-making (and resistance to it) through Neferti Tadiar's conception of life-times as a frame to foreground "the texture and qualities of experience and times of living" that are enrolled in creating value for the utility.[13] The creative, affective, familial, and life-making strategies that people practice to sustain their households—and afford electric bills—become a resource for the utility. The state authorizes a private company to wager the resourcefulness of working-class, unemployed, underemployed, elderly, and fixed-income customers who find a way to make do. This burden on low-income households becomes even more stark given the continued underfunding of means-tested relief programs to reduce utility bills in Georgia. In contrast to other states where ratepayers fund utility bill discounts for income-qualified households,[14] only a small discount for seniors, public housing residents, and disabled customers and limited federal funding for seasonal bill payment assistance exist to help customers with their utility bills.[15] This is despite significant need in Atlanta and across the South more broadly, where households report the highest rates of energy insecurity in the US; 30 percent of households in the region reported experiencing energy insecurity in 2020, defined as difficulty paying energy bills or keeping their home at an unsafe temperature to limit energy costs, and 14 percent reported receiving a disconnection notice.[16] Employing the lens of feminist urban political ecology, I view electricity pricing through the time spent on household reproduction captured by the utility and contrast this with campaigns in which customers demanded a right to affordable energy as a basic need in a call to democratize and decommodify electric life.

Inducing Demand in the Reproductive Geographies of Electric Life

In the early twentieth century, as Georgia Power's influence as a quasi-state development entity was consolidated, the utility initiated a campaign to increase power sales to residential customers. Electrification reworked intimate geographies of cities by domesticating "good nature" that "became part of (and a basic precondition for the construction of) the protected inside of the modern home."[17] In a speech before the Edison Electric Institute, Georgia Power president Preston Arkwright prided the utility on providing the first electric power many had used as the utility entered "intimately into all the life of the community, into all the homes and all the stores and all the industries and all the activities."[18] These intimate circuits imply a personal relationship between the utility as a community member, a citizen, and a welcome visitor in the interior and intimate spaces of electric life. The intimacies of the electric grid that connect "differentially laboring peoples" rewired the time of social reproduction and created new opportunities for electricity capital as the need for electricity and the hundreds of consumer goods it powers mushroomed.[19]

Household technologies that promised increased efficiencies in reproductive labor proliferated through financing and marketing by the utility.[20] As early as 1911, Georgia Power's predecessor, Georgia Railway and Power Company ran promotional holiday campaigns for lights, irons, toasters, fans, and percolators.[21] Beginning in World War I, as women entered paid employment in clerical or war work in greater numbers, many struggled to hire domestic workers, leading electric companies to suggest in advertisements "that housewives consider 'electric appliances as their servants.'"[22] The anthropomorphic electric bolt Reddy Kilowatt, first drawn by an Alabama Power Company illustrator, was meant to assuage women's fears of using electric appliances.[23] This "willing servant who works long hours at low wages" could emancipate women from wearying household reproductive labor and enable more efficient time management.[24] An electric range could save fourteen eight-hour days over years, promising the ability to "produce more, in less time if it's ELECTRIC." A refrigerator could save "time and money," up to eight and a quarter eight-hour days over a year as women's work shopping for, cooking, and preserving food was replaced by kilowatt-hours.[25] By replacing the time of household labor with electric power, the utility promised women could not only care more effectively for their families but also "turn back the clock," because "drudgery, crushing out youth, began to vanish as woman's penalty for her sex when the electric light was first introduced into the home. . . . Electric service has created a brand-new environment for all of us, and of all its benefits, womankind has reaped the major share—in increased years of youth!"[26] Electric life extended the "tyranny of time" as the "normative measure for activity" into all aspects of life's work for the homemaker and those whom she employed as new technologies for cooking, cleaning, and caring for children industrialized household labor.[27] Advertisements to induce demand sold an image of white feminine modernity powered by gadgets as well as the private model of power provision, boasting of Georgia Power's service, in opposition to the "deadly threat" of "government ownership" (see advertisements in figures 2.1 and 2.2).[28]

Electricity transformed the intertwined geographies of paid and unpaid reproductive labor. Georgia Power's home economics team aggressively marketed electrical appliances. Mobile electric kitchens staffed by "husband-and-wife teams" toured the state to teach women how to use new technologies and demonstrate "how electric living was better."[29] A parallel track of classes trained domestic workers "in the proper uses and care of electric appliances and equipment."[30] New appliances also returned work to the home that people had previously paid to have done by others, disrupting the waged laundry and domestic labor industries almost entirely, although not exclusively, staffed by Black women in Atlanta.[31] In 1930, while a little over half of Black women worked, 90 percent were employed in laundry or domestic service, often for white households, and continued to work in large numbers in domestic service until the civil rights movement. The electrification of care work did not deliver on the promise of erasing demands of tedious and underpaid household labor or address segregation in the labor market that limited work available to Black women and women of color and devalued their time.[32] Rather, electrification enabled a double shift that created new obligations for unpaid reproductive work in the home even as paid care work continues to be devalued.[33]

Figure 2.1
Georgia Power advertisement, 1933. Photo Credit: Southern Israelite, Georgia Historic Newspapers.

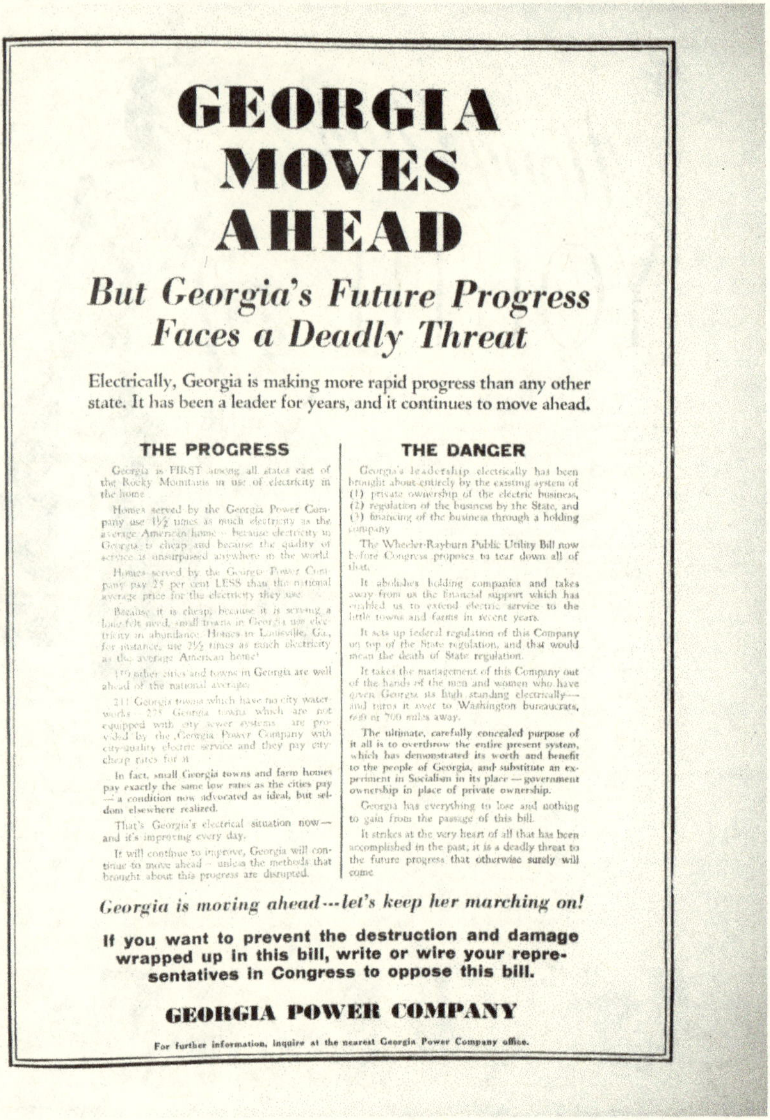

Figure 2.2
Georgia Power advertisement, 1935. Photo Credit: Southern Israelite, Georgia Historic Newspapers.

Residential electrification received a boost with New Deal policies that also contributed to making electricity ubiquitous in homes.[34] The Federal Housing Administration instituted uniform guidelines on electrical outlets and required that homes purchased with an FHA-backed mortgage have sufficient circuits to handle electric use from appliances and lights.[35] Georgia Power advertised appliances financed through the Electric Home and Farm Authority, using the program more than any other private utility.[36] Between 1927 and 1942, kilowatt-hours sold per customer tripled, as shown in figure 2.3.[37] Electricity use spiked again with affordable air conditioning window units introduced in 1951, although they remained a privilege in the first decades they were commercially available.[38] As residential demand for power increased, so did total company sales.

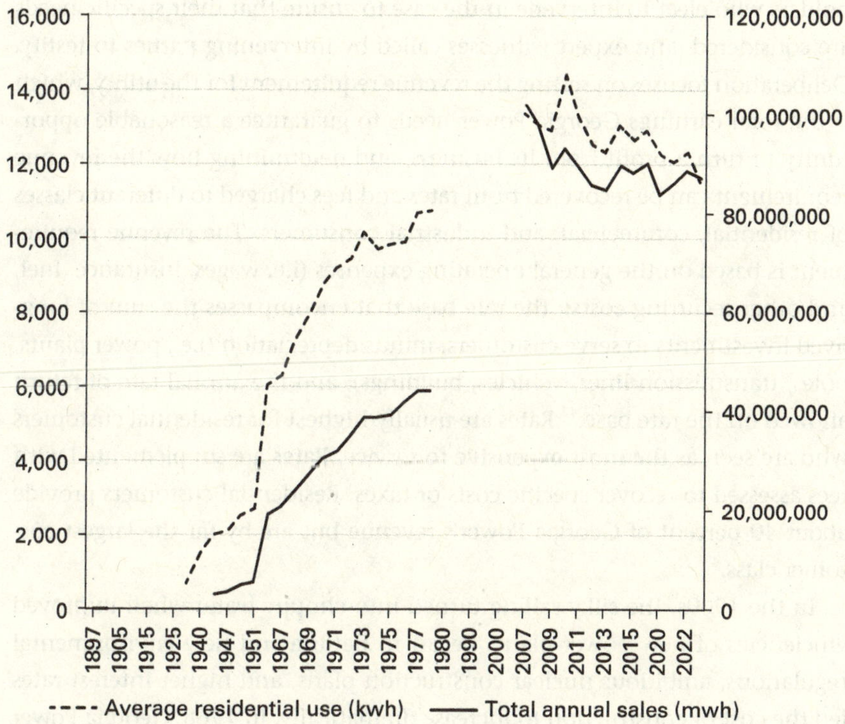

Figure 2.3

Increase in average residential demand for Georgia Power customers and increase in total company sales, 1897–2022.[39] Graph by author.

The electrification of everyday life depended on collaboration with the Commission to build power plants to meet growing demand. In the era after World War II, technological improvements enabled new and larger coal-fired power plant construction.[40] The Southern Company attributes its "declining-cost business and silky sailing through postwar America's glory days" to these technological efficiencies as the real price of electricity decreased 71 percent between 1940 and 1970. Consumption increased by "leaps and bounds," doubling each decade as people came to depend on electricity for the mundane routines of household reproduction. The utility went before the Public Service Commission to request only two rate increases and ten rate reductions.[41]

The process of adjudicating rates is a complicated legal and financial determination. The Commission reviews testimony from the utility, stakeholders who elect to intervene in the case to ensure that their specific needs are considered, and expert witnesses called by intervening parties to testify. Deliberation focuses on setting the revenue requirement for the utility, which is the total earnings Georgia Power needs to guarantee a reasonable opportunity to turn a profit from its business, and determining how the revenue requirement can be recovered from rates and fees charged to different classes of residential, commercial, and industrial consumers. The revenue requirement is based on the general operating expenses (i.e. wages, insurance, fuel, and other recurring costs); the rate base that encompasses the sum of long-lived investments to serve customers, minus depreciation (i.e., power plants, poles, transmission lines, vehicles, buildings); and the annual rate of return allowed on the rate base.[42] Rates are usually highest for residential customers who are seen as the most expensive to service. Rates are supplemented with fees assessed to recover specific costs or taxes. Residential customers provide about 40 percent of Georgia Power's revenue but are by far the largest customer class.[43]

In the 1960s, the silky sailing turned into choppy water when improved efficiencies of new power plants began to flatline and new environmental regulations, ambitious nuclear construction plans, and higher interest rates led the cost of construction to increase dramatically. In 1968, Georgia Power requested "rate relief" from the Commission and was allowed "an upward adjustment" of 2.172 percent on rates across all customer classes. Large industrial customers, from the US Army to US Steel, and residents spoke out against the increase, outraged that the utility sought to pass on the costs of inflation.[44] The hearings sparked the beginning of sustained opposition as the utility returned to the Commission annually to request rate increases. Nascent environmental and consumer rights activists, the Atlanta Chapter of the Welfare Rights Organization, the American Association of Retired Persons (AARP), labor unions, and public officials would all speak out against the tax the utility extracted from electric life with the Georgia Power Project launching an extended campaign to politicize the rate hikes.[45]

Consumer Advocacy of the Georgia Power Project

In the 1970s, the Georgia Power Project organized to challenge Georgia Power and protest the organization of the utility that took advantage of working-class people in Atlanta. The 1968 rate request drew ire in the pages of *The Great Speckled Bird. The Bird* was a radical newspaper founded by members of the Southern Student Organizing Committee who had convened in Atlanta to support the civil rights and antiwar movements and covered wide-ranging local and national issues including feminism, gay liberation, environmentalism, and energy politics.[46] In coverage of rate cases between 1968 and 1985, writers at *The Bird* protested the utility's control over electric life that transformed electrical power into political power, or "the ability to force people to do your will." Atlantans suffer the increases and "pay our electric bill, forgetting that $.20 per month, because we need the power and there is no where [sic] else to get it."[47] Writers and organizers working with *The Bird* established the Georgia Power Project in 1972 after Georgia Power filed its third rate request in four years.[48]

The critiques levied by the Georgia Power Project were formalized in eight demands that spoke to a broader coalition representing working people's everyday concerns (see figure 2.4).[49] These offered tangible strategies to reign in the financial model of the utility "as a public utility run for private profit" and limit its hold over the times of electric life from restricting deposits and shutoffs to improving energy efficiency.[50] By organizing around "pocketbook issues," the Georgia Power Project explicitly endeavored to unite "diverse groups (labor, community, radical, ecology, consumer, poor, etc.) against a common corporate Enemy."[51] Although there had been instances of consumer organizing throughout the Commission's history, including the Atlanta Chapter of the Welfare Rights Organization and the Atlanta Domestic Worker's Union petition against the 1971 increase,[52] "there was no real residential

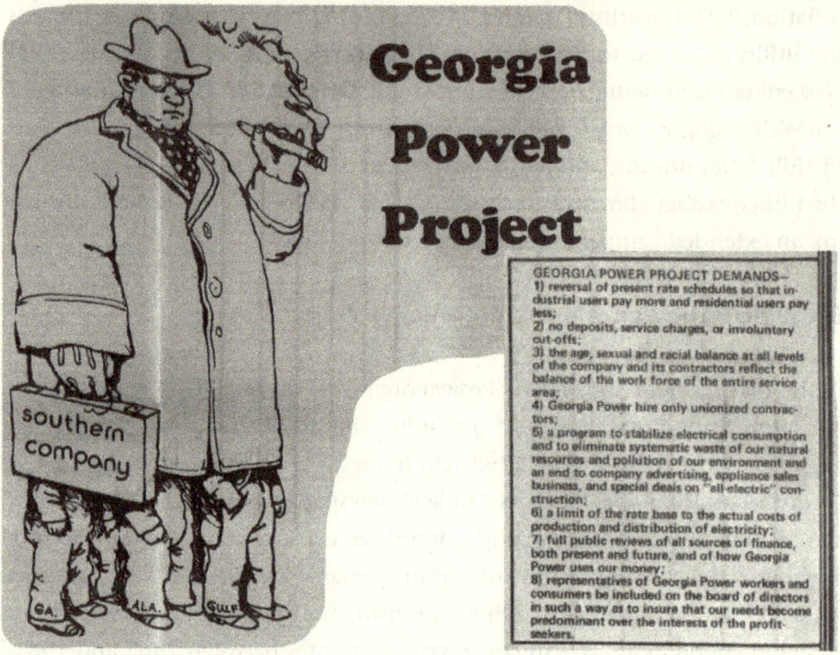

Figure 2.4
Georgia Power project demands. Image credit: Copyright Atlanta Cooperative News Project, courtesy of Special Collections and Archives, Georgia State University Library.

agitation," explained Liam, a former Georgia Power Project leader.[53] Notably, these demands have endured as consistent requests that consumer advocates have raised repeatedly in the half century since.

The Georgia Power Project utilized a variety of tactics to push the Commissioners to take a more active role in regulating the company. They traveled across the state to sites where Georgia Power had proposed to build large new power plants, including Plant Scherer and Plant Vogtle Units 1–4, gaining 600 members statewide by 1975.[54] In Atlanta, they partnered with the electric workers and industrial unions, picketing with the Atlanta Labor Council outside of Commission hearings, recruiting participants to testify at hearings, petitioning to hold hearings after working hours to allow more people to testify, challenging the utility's television advertising with the Federal Communication Commission under the fairness doctrine, and intervening in the legal process of the rate case.[55] By participating in regulatory discussions, the Georgia Power Project disrupted the utility's monopoly on expertise while public tactics helped win press coverage of the otherwise obscure bureaucratic functions of the Commission.

Efforts to rally public opposition won attention from elected officials, including then Georgia governor Jimmy Carter and Commissioner Bobby Pafford. The Commission allowed the utility to collect only a portion of the requested increase in 1972 and initiated a study of rate structures to evaluate the distribution of the revenue requirement between customer classes that the Georgia Power Project had encouraged.[56] The Georgia Power Project also submitted a proposal to end declining block rate structures in which the price per kilowatt-hour fell with increasing use, arguing this rate schedule discriminated against the smallest users and should be reversed to encourage efficiency and to require larger users who are more able to afford electricity to pay more. In addition, they sought to end differences in rates charged between different customer classes that gave industrial customers a discount relative to households.[57] The Commission ultimately required only small adjustments across rate classes in 1973. As a result, the Georgia Power Project sued the utility over its rate increases. The State Supreme Court eventually ruled that consumers may not sue over the cost of power being too high.[58] In another case, the Georgia Power Project and *The Great Speckled Bird* sued Georgia Power after power to its office was disconnected. They argued the shutoff was retaliatory and that electricity, which is "a necessity of urban life," should be disconnected only after a hearing, especially given that "the Power Company operates under strong state regulation, its actions are effectively those of the state" and thus subject to due process.[59]

Although these suits were unsuccessful, the Georgia Power Project effectively called into question whether the utility was serving the public interest. Gregory, a former Georgia Power Project member, described that their research exposed how the utility manipulated the public process "to extract money. . . . [So our work] was both this kind of philosophical commitment to an anti-capitalist culture and spreading the gospel of the public good" in a theoretical and real way because "we were in the weeds, demonstrating what it actually meant."[60] Members of the Georgia Power Project studied utility history and finance and came to critique the foundational basis of the utility's rate requests, which assumed "exponential growth."[61] Georgia Power had manufactured enormous increases in power demand through advertising and marketing of appliances and industrial development promotions.

Although this demand had begun to plateau with the economic downturn in the 1970s, the utility continued to plow ahead with an ambitious $1.5 billion construction plan.[62]

The Georgia Power Project criticized new construction that they argued was unnecessary except to make profits. Roger recalled in an oral history, that they tried to figure out why Georgia Power was selling bonds to finance construction when "its capital structure was getting way out of whack with too much debt, not enough equity. So the bond prices were going up, and the reason of course was that all the profits were going to build the nuclear plant, so their profit level was coming down." For Georgia Power, this threatened either default on their bonds or not being able to send returns to Southern Company to provide dividends on the holding company's stock. While profit had increased for the power company with increases in efficiency at larger plants, the company eventually saw diminishing returns to scale as new nuclear and coal plant construction added additional costs.[63] Yet, the utility's business model depending on building, even when construction was uneconomical. Liam argued, Georgia Power was "so addicted to building plants. . . . They were a bank. The power business was an excuse. . . . The power bank sold electricity, and they built power plants—that's where the money was."[64] The utility argued that without this investment blackouts loomed, and the state would be "short of energy for its needed recovery . . . [cutbacks in the construction budget] would only worsen the economic conditions now prevailing."[65] In contrast, the Georgia Power Project surmised that more expensive construction, combined with inflationary pressures that increased the cost of capital, meant "there were no new profits . . . to be gotten from the electric generation business itself and they had to raise rates to extract profits from the other parts of the economy."[66] In a joint press conference with the United Auto Workers, Jane Switek with the Georgia Power Project asserted that the Commission should deny any further rate increases and "act to solve the real emergency of the people of Georgia, not Georgia Power's sham emergency."[67] What was at risk was not reliability (after all, Atlantans were already having to manage brownouts while Georgia Power sold power out of state[68]) but returns pegged to capitalization for the utility.

Critics of the utility argued that constant construction left Georgia in a dependent position "subject to the will of the Yankee money markets," because of "the tremendous amounts of capital" that the utility required to fuel its growth.[69] As Manning Marable summarized a decade later, "Southern

Company's influential investors are from Wall Street and reflect finance capitalism's direct relationship with the economic development of the colonial economy of the South."[70] In 1974, after Georgia Power failed to secure its full rate request from the Commission, Southern Company sold stock to an underwriting syndicate managed by Morgan Stanley, Merrill Lynch, and Saloman Brothers.[71] The quick infusion of cash saved part of the construction program as the utility wagered future demand for power would pick up. Given its limited cash flow, Georgia Power ultimately canceled, suspended, or postponed construction on eighteen generating units.[72]

To break the link between financial and electricity capital would require not only an aggressive regulatory role for the state but a change in the institutional configuration of governance to reign in the company's ability to speculate on electric life. Columnists debated proposed alternatives in *The Bird*. Bob Hall with the Institute for Southern Studies argued that TVA and other government-owned utilities still depended on the private bond market, and "non-profit, state ownership of Georgia Power" would be insufficient to achieve independence. Instead, democratizing "control and ownership of the utility by a council of its workers and users" could shift the utility so that "policy would be determined by working people instead of by bankers and capitalists."[73] Another author encouraged municipalization through a public takeover by the City of Atlanta.[74] At the same time that Atlanta advocates debated public power alternatives, economists who would come to be associated with neoliberalism began to attack the regulated utility model, arguing market mechanisms could assure greater efficiency and lower prices through competition.[75]

Facing myriad calls for reform, Georgia Power supported the Territorial Electric Service Act that eventually passed the General Assembly in 1973. Intended to introduce limited competition to quiet antitrust challenges to the utility's control over nuclear power production, the law stipulated that new industrial or commercial customers with anticipated load at or above 900 kW could select their utility provider. It also permanently fixed in place utility service boundaries.[76] The Georgia Municipal Association, Atlanta Board of Aldermen, and Atlanta Labor Council opposed this move for deeding undeveloped land to either electric co-ops or Georgia Power, effectively foreclosing the expansion of any new public utilities by preventing "a city not presently in the electric business to decide to buy its own power system at any time in the future."[77] The act located Atlanta within the boundaries of Georgia Power's service territory and remains a barrier to the city trying to procure energy resources from any other utility.

Negotiation over the bill also laid the groundwork for Georgia Power's partnership with electric co-ops and municipal utilities that ultimately saved its construction program. Oglethorpe Electric Membership Corporation, a bulk power purchasing cooperative formed by thirty-nine electric membership cooperatives that had previously purchased power from Georgia Power without owning generation capacity, offered to purchase a stake in Georgia Power's power plants that were still under construction with low-cost federal loans it could access through the Rural Electrification Administration. With this deal on the table, Georgia Power reversed their competitive stance with the co-ops; company president Bob Scherer described the co-ops not as enemies but "as deep pockets."[78] While the Southern Company history attests that "the sale of generation to federally backed cooperatives flew in the face of Southern Company tradition," in practice, the arrangement represented a new manifestation of the symbiotic relationship between the utility and state.[79] In his history of Southern electrification, Cater argues investor-owned utilities consistently "transformed government—at the local, state, and federal levels—into a protective shield, a generous benefactor, and a channel through which private energy corporations could capture public resources."[80] The infusion of public capital assured the completion of the Plant Hatch and Plant Vogtle Units 1 and 2 and Plants Bowen and Scherer as Georgia Power stayed the course on growing and building.[81]

In building solidarity around pocketbook issues, the Georgia Power Project laid the groundwork for organizing to protest Georgia Power and politicize rate-making at the Commission, even after the group dissolved. Popular organizing led the General Assembly to create the independent Office of Consumers' Utility Counsel in 1975 to represent residential customers and small businesses before the Commission.[82] Unfortunately, the Office was short-lived. It was moved to the Governor's Consumer Protection office in 1995, and eventually defunded in 2008, just before Georgia Power received approval from the Commission to build Units 3 and 4 at Plant Vogtle.[83]

Construction Work in Progress

The persistence of the grow-and-build model enabled by the state is most obvious in the history of Plant Vogtle. When construction began in 1976,

Georgia Power faced opposition from the Georgia Power Project, environmentalists, and the co-ops and municipal utilities. While partnerships with the co-ops and municipal utilities eventually rescued the project in a time when Georgia Power feared its own financial solvency, Vogtle continued to face cost overruns that again threatened the utility's financial position. Novel financing mechanisms, enabled by the state and federal governments, secured Georgia Power's nuclear program at the expense of captive ratepayers.

In the 1980s, Georgia Power revived a suggestion it had first proposed a decade prior to secure additional revenue by charging customers for "construction work in progress." The utility argued that customers would benefit from paying for construction along the way, rather than paying off the balance of construction plus interest after the plant's completion. Charging for work in progress goes against the traditional practice of ratemaking in which the company assumes risk and customers pay only for capital assets in service from which they directly benefit.[84] Construction work in progress converted customer's future demand for energy into a resource that could be advanced as a revenue stream in the present.[85] In 1985, the General Assembly convened a study to examine phasing Vogtle into the rate base—an alternative formulation of construction work in progress. Opponents sought to cap the total costs that the utility could recover, asserting that Georgia Power's debt was the result of poor planning, the costly plant was not needed, and the utility had refused to consider cheaper alternatives, including cogeneration.[86] Amid opposition, the construction-work-in-progress study prepared for the Joint Comprehensive Energy Resources Committee was shelved in 1987, the year the first Vogtle reactor came online.[87]

Debts associated with Vogtle presented a liability to the utility as deregulation took off across the country. The Public Utility Regulatory Policies Act of 1978 had introduced the first measures creating a market for electricity generation. These were expanded with the Energy Policy Act of 1992, which allowed states to introduce competition for wholesale and retail power.[88] Georgia Power petitioned the Commission to allow accelerated depreciation to quickly pay down Vogtle Units 1 and 2, arguing that long-term, higher rates tied to paying off the debts of nuclear plant construction might be less attractive to footloose customers seeking the lowest cost electricity as states experimented with deregulation.[89] The request raised excess cash from ratepayers and coincided with a purchasing spree at Southern Company. Georgia

Power's parent company bought power plants in California, New York, and Massachusetts, as well as South Western Electric Board in Bristol in the United Kingdom and Consolidated Electric Power of Asia.[90] Ultimately, Southern's merchant power business, Mirant, was spun off from the traditional utility business. Mirant went bankrupt following the Enron scandal, while Southern and its holding companies turned to lobby against deregulation.[91] A Southern Company executive remarked that after studying the effects for customers in different state's energy markets, they found deregulation introduced risk without "oversight over the entire system to make sure it works and make sure it's reliable."[92] Monopoly control over the grid is an "old, conservative tried-and-true strategy"[93] that offered reliability, both for the grid and for shareholders who were guaranteed returns.

Pivoting away from deregulation, Southern Company redeployed its tried-and-true grow-and-build strategy. The same month the Energy Policy Act of 2005 was signed into law to create the DOE Loan Program Office, Southern declared their intent to investigate nuclear expansion with two new reactors at Plant Vogtle. Vogtle reactors 3 and 4 would be the first new nuclear construction permitted in the US since the Three Mile Island accident in 1978. Southern Company applied for and received $8.3 billion in federal loan guarantees for Vogtle intended to spark a nuclear "renaissance" in the US.[94]

In addition to US taxpayer money from the Loan Program Office, the utility preemptively sought advanced ratepayer support through a construction work in progress charge.[95] The Georgia Nuclear Energy Financing Act adopted by the General Assembly in 2009 with the endorsement of the Commission allowed Georgia Power to collect a nuclear construction cost recovery (NCCR) charge for work beginning in 2011.[96] The fixed fee added almost $100 a year to customer bills to generate approximately $500 million annually.[97] Both the loan guarantees and the NCCR "transfer financial risk away from Southern Co. and its shareholders and toward consumers and taxpayers," cautioned the *Atlanta Journal-Constitution* in 2012. Then CEO Tom Fanning admitted that this regulatory and financial support was a necessity: "There's no project like this that would be built without that kind of support."[98] In contrast, the advocacy group Nuclear Watch South argued the NCCR "removes the liability of speculation on financially risky reactor investment from the electric companies and places the risk on residential and small business electric customers" who are forced "to become investors" in future plants, and forced to make changes to everyday life to afford higher bills.[99] Elderly people without a guarantee that they would ever be able to see the benefit of new generation are hurt the most. In practice, customers did not receive the benefits of "investors" who might see a return on their investment through dividends or interest on bonds as the NCCR taxed everyday life—extracted in monthly increments on power bills.[100] Yet, as Tadiar observes, the "seemingly limitless resource that is the future (as part of the seeming limitlessness of life itself) is in actuality the lives of people whose own futures are offered up as exchange values extractable in the present."[101] Georgia Power cashed in the future electric demand of its customers to raise capital for nuclear construction based on the understanding that ratepayers in a captive territory will continue to use nuclear-powered electricity for years in the future.

The NCCR, which was scaled as a percent of a customer's base bill, added to an environmental charge for pollution-control technologies; a fuel charge, which includes the variable costs of coal, gas, nuclear and biomass fuels; and up to nine other fixed charges. The use of fixed fees to recover costs from all customer bills—rather than rates that are scaled to kilowatt-hour use—further exacerbated already uneven energy burden.[102] Hugh, an energy consultant, described that "in Georgia Power's zeal to sell Plant Vogtle and to position itself as a great utility and . . . because the Public Service Commission is elected, rather than appointed, public service commissioners obviously have an incentive to try to paint electricity rates in Georgia as being lower than the national average. Well, there's a little bit of smoke and mirrors that's going on here."[103] While the volumetric rates have remained low, when accounting for the full electricity bill, in recent years, Georgia Power customers have paid some of the highest monthly costs in the country.[104] Income inequality in Atlanta further contributes to the city having the fourth highest share of low-income households grappling with unaffordable energy burden in the US.[105]

Commission support for Vogtle continued, even after its sister project, VC Summer in South Carolina, was discontinued and costs began to balloon.[106] When Vogtle's expansion began, the total cost was anticipated to be $14 billion, and the plant was expected to come online in 2016. Learning from the cost overruns of Units 1 and 2, the Commission established biannual reviews of the project and hired an independent monitor to provide additional oversight. Nonetheless, delays plagued the project and eventually led to the bankruptcy of the contractor, Westinghouse. A Southern Company subsidiary,

Southern Nuclear, took over management. In December 2017, the Commission required Georgia Power to issue a one-time $75 credit for all residential bills as part of its authorization for work to continue.[107] When this decision was announced, the then Commission chairman Stan Wise explained that the Commission did not go further to penalize the company because "the long-term benefit that we will receive over the life of this project was the overriding factor."[108] Three months later, the Commission reduced the NCCR fee by $1.03 a month to keep "more money in the pockets of Georgia Power customers."[109] It also reduced the revenue requirement of the utility, capping the total cost of Vogtle that could be charged to customers and forcing the utility and its shareholders to shoulder the remaining debt. The projected "long-term benefit" of Vogtle is widely debated. Former secretary of energy under President Obama and later Southern Company board member Ernest Moniz asserted the importance of the project to address climate change.[110] Secretary of energy under the Trump administration Rick Perry also lauded the project as the "real New Green Deal," in mocking reference the Green New Deal, when announcing an additional $3.7 billion in loan guarantees in 2019.[111]

When Unit 3 finally came online to much fanfare for the nuclear industry in 2023, reactions in Georgia were more mixed as the full cost of Vogtle could now be added to customer's monthly power bills. The Commission ultimately permitted Georgia Power to collect $7.6 billion in remaining construction costs. Shareholders will pay $10.2 billion, with the balance split between co-op and municipal electric utility co-owners. The average residential customer saw a $9 a month increase in 2024, adding to the $1,000 the average household has already paid for Vogtle—double the anticipated cost of the NCCR if construction had been completed on time. As part of this negotiated settlement, Georgia Power agreed to expand its Income-Qualified Senior Citizens Discount program, available to customers sixty-five years of age or older with income equal to or less than 200 percent of the federal poverty level, to include households receiving Social Security Disability assistance, Supplemental Security Income, or a Section 8 housing voucher. The maximum value of the monthly credit also increased from $14 to $33.50.[112] The decision recognizes a deserving poor who may be harmed by rising energy costs, while refusing to acknowledge how Vogtle stresses household budgets for many more households in Atlanta.

Beyond the meager bill discounts from the utility, public assistance in Georgia is limited to charity programs from the Salvation Army, churches, and federal funding for the Low-Income Housing Energy Assistance Program (LIHEAP) that is administered by the Fulton Atlanta Community Action Agency from December–March for heating and May–August for cooling. LIHEAP is not a universal program. States are allocated a limited pool of formula funds, which are awarded on a first come, first serve basis to applicants who complete a multistep process. On average, LIHEAP only meets about 20 percent of estimated need in Atlanta. And, explained one researcher, getting assistance "does require an extra step of being more engaged. . . . It does require more work."[113] This could be repeated trips to resource centers, as one mother told the *Atlanta Journal-Constitution* while waiting outside with her six-year old to be seen at the Fulton Atlanta Community Action Authority, hoping "I'll be able to get in."[114] Nonprofit providers also visit apartment complexes and retirement communities, taking on social reproductive work that is outsourced outside the utility to ensure continuity of service for poor customers.[115]

Maintaining shareholder value calculated from the capitalized investments of the utility that spiked with Vogtle comes at a steep cost for poor people who are disproportionately people of color in Atlanta. The refusal to acknowledge or address the regressive effects of rates must be understood within the devaluation of labor and life-time of non-white people. Rebecca, an energy advocate argued that the acceptance of painfully high energy burden in Atlanta

> goes back to the foundation of what the Southern economy [is] built off of. So, you have structural racism and inequities that don't allow for a true desire to serve everybody who lives here. There's this belief that haves and have-nots are the right thing . . . that people should just pull themselves up by their bootstraps, and it's America so everybody's equal and there's this fallacy of equal access to education to housing, on and on. If you're coming in with that belief, then it's going to be really hard for me to talk to you about how bad and inadequate the housing stock is in Atlanta for the working poor.[116]

Commissioners coming from this perspective blame poor people who struggle to afford their energy bills for personal failings, refusing to see the systemic constraints utility regulation—itself shaped through white supremacy, racial capitalism, ableism, and sexism—exacerbates for poor people who must work a second job, stand in line at churches or community centers to receive public assistance, or ration energy, sweating in the summer heat and layering sweaters in the winter, to keep their light bill affordable. This necessary time, effort, and worry that people spend to pay the cost of electric life does not receive a wage and is disregarded by the Commission, even as the utility is able to collateralize these ordinary demands for energy in its protected service territory, wagering the lengths to which people go to afford their power bills. Affordability is the overwhelming concern for the majority of Atlantans who pay residential electric bills. Yet, in proceedings around Georgia Power rates, the everyday experience of electric life pales in comparison to other priorities for commissioners, who see themselves as regulating to advance the financial health of the utility rather than representing Georgia residents.[117]

Fight the Hike for Fair Energy Now!

In June 2019, Georgia Power submitted a proposed rate case to the Commission that called for an increase in rates for the first time since 2013. This increase would allow the utility to recover an additional $2.2 billion over three years to fund grid improvements, replenish its natural disaster restoration fund, and clean up coal ash waste. Plant Vogtle was not mentioned but loomed in the background of the proceedings. The proposal would have added $16.48 a month to the average residential bill through both an increase in the rates per kilowatt-hour and an increase in the basic monthly service fee charged to all residential customers before they turn on their lights from $10 to $17.95.[118] The proposed increase of the service fee became a rallying point for community organizing to Fight the Hike.

The Fight the Hike campaign united consumer, faith, and environmental organizations led by Black women, including Sankofa Church, Mount Welcome Missionary Baptist Church, Malcolm X Grassroots Movement, Pan African Orthodox Christian Church, SisterCARE Alliance, and the Georgia Conservation Voters. The coalition hosted town halls across the state, gathered letters in opposition to the fee hike, phone-banked to turn people out to the monthly hearings on the rate case, organized neighborhood carpools to hearings, and hosted two rallies outside the Commission building in downtown Atlanta where working people, students, elders, and church members who were not able to attend the Monday morning hearings gathered.[119] The Georgia Chapter of the Sierra Club and Georgia Solar Energy Industries Association also hosted community meetings in the districts of each commissioner.[120]

When hearings began on the rate case, coalition members packed the Commission hearing room for public comment. The hearing room is shaped like a courtroom. Company lawyers sit on the right side and Commission staff sit on the left side behind the court reporter, facing a raised dais where the five commissioners sit. Industry, environmental, and public interest lawyers and members of the public occupy four cramped rows of black leather chairs at the back. On the first day of months of hearings, Chairman Lauren "Bubba" McDonald called the room to order and invited testimony from then Georgia Power CEO Paul Bowers, who declared the proposed rate increase was essential to maintain the flows of electricity that are a "cornerstone of Georgia's economy." Emphasizing the foundational role of regulation for the utility, Bowers thanked the Commission for its past work that gives "businesses and industries the confidence to locate in our state and have given investors the confidence to provide the capital necessary to finance the infrastructure that is required now and, in the future."[121] Others would go on to speak in favor of the rate increase, including members of the International Brotherhood of Electrical Workers union, companies that partner with the utility, and a religious charity with a Georgia Power executive on their board.

Yet, as one person identified after a month of the proceedings, "Most of the public people [who] are not paid per hour to be here, as many of the professionals are," came to speak against the increase.[122] Petitioners shared a range of reasons for opposing the hike that contested how an increase in electric bills would stress their time—as one former city employee described, "I am also proud to be retired, and want to stay retired."[123] Another elderly woman invoked her lifelong contributions to the company as a Georgia Power ratepayer: "I'm a Grady baby. I've been in Atlanta all my life, and my family has supported Georgia Power all our life, and I really appreciate the service that Georgia Power provides us. But I want to ask you that you stop the increase because I can't afford it."[124] A Savannah resident described working with a community organization, the Harambe House, to gather 1,400 signatures opposing the hike, stating, "I also run a small Medicare enrollment business, and I drive Pizza Hut delivery on the weekends, and I am a fulltime mother of one. . . . I myself cannot afford the rate hike with doing this work." While collecting signatures, she met many of her neighbors who were forced to switch to a prepay meter after a disconnection. The meter charges on a pay-as-you-go basis, billing the account for each kilowatt-hour used until the account balance reaches zero. She explained: "One parent told me that . . . she didn't even have $5 to her account for her kids to even have power to get dressed for school. I can't imagine being a child in America that is getting dressed in the dark, and the rate hike would only cause more people to be in that situation."[125] Prepay meters extend the utility's control over time, imposing new stresses on residents to economize each action that uses electricity with the ever-present threat of disconnection.

Many asked the commissioners to consider that energy costs do not exist in a vacuum. Two Spelman graduates testified to the compounding tax that a fee hike would impose on their daily lives made more difficult by a discriminatory labor market. One woman several years out of college explained that after years away, she returned to Atlanta where a fruitless job search left her working "low-wage jobs ranging from $7.25 an hour to $15 an hour."[126] Another alumna a few decades out of school echoed that

> due to an on-the-job injury, working a low wage job because I, too, had to take what I could to try and keep the lights on, I am now also a resident on a fixed income making $15,000 a year. . . . With record high and low temperatures in recent years, we, the consumers, have had to endure unusual spikes in usage costs for those of us that are at or below the federal poverty level. We have also had to forgo some basic normalcy. It is not uncommon for us to have to choose between air conditioning or opening windows during the hottest and most humid days or not turning on the heater to stay warm on the coldest days and nights, instead relying on the stove and oven to warm our homes while we cook our meals.

She admitted, breaking into tears, the stress she faces as her time is spent on basic social reproduction:

> It is a continuing challenge to juggle payments on the electric bills to keep from having the service cut off altogether. We all know the massive financial hardship that comes into play if your utilities are cut off: payment in full, required; additional deposit, required; adverse impact on your credit score, and in many cases that involve renting if your utilities are turned off, you are considered in breach of contract and risk being evicted. So, you see, for me, and I am sure I am not alone in this. It is a very tenuous situation. Just one instance of not paying an electricity bill in time will begin a domino effect that I fear may result in my becoming homeless.

She turned her plight back to the Commission, asking, "Did you ever have to opt not to seek needed medical attention because your electricity bill had to be paid or find yourself meticulously monitoring your daily power usage?"[127] She implored the commissioners to remember why this vital commodity is not left to the market alone and act on their mandate to maintain reliable and reasonably priced service for all Georgia residents, rather than abetting the extraction of value from the ordinary moments of electric life.[128]

In December, when the Commission returned to hear closing arguments, the room was decorated for the holidays with artificial holly pinned to the dais and tiny wreaths with red velvet bows hanging below each commissioner's nameplate. Coalition members and others filled the back rows of the hearing room wearing red, not to signify holiday cheer but opposition to the hike. Chairman McDonald called the meeting to order and announced that a stipulation to the proposed rate case had been filed by Georgia Power in agreement with the Georgia Industrial Group, the Commercial Group, the Georgia Association of Manufacturers, Kroger, the Metropolitan Atlanta Rapid Transit Authority, and the City of Atlanta.[129] Intervenors who represented these and other parties were called to present their final opinions before the Commission would issue its decision. Many spoke against the stipulation or to issues that had not yet been resolved. The Public Interest Advocacy staff for the Commission, who are charged with reviewing the filings and making a recommendation to the Commissioners on behalf of the Georgia citizens, explained why their office had not signed on to the stipulation brokered between the companies and its biggest customers.[130] A lawyer expressed that there was "no doubt that the parties who have signed on believe they're getting a benefit with result with respect to rate design. . . . It certainly appears that in return for that favorable rate treatment, the parties have agreed to a [capital] structure, particularly the earnings band that would provide ratepayers with significantly less benefit."[131] In other words, he continued, the largest users negotiated favorable electric rates while cosigning higher residential costs and increased returns for the utility. Under this agreement, Georgia Power could make up to $330 million a year above the national average for investor-owned electric utilities before being asked to return any of that profit to customers. The Commission attorney called into question the utility's commitment to the public, noting that "the company testified in this case that the interests of shareholders and ratepayers are aligned." If the interests are the same, he continued, "it shouldn't take over three hundred million dollars a year to incentivize the company to act in the customers' best interest."[132] In closing, he emphasized that the elected Commission could reject the agreement between Georgia Power and its largest customers and determine a rate structure that would secure more affordable electricity for all.

A Georgia Power representative responded, and the floor was open for comment from the public. The executive director of Georgia Women for a New Direction was the last person to speak. She reframed the points that the Commission staff raised as a value proposition:

> Living in a state that is part of a region that already has and will continue to bear the economic and environmental burdens of climate change, it is unconscionable to think that you all, our public service commissioners, would knowingly decide to pour gasoline on the fire and require Georgia residents to pay even more to be impacted in a worse way. . . . Your decisions show in practice the values of the Public Service Commission—some people are more valuable than others. The Public Service Commission has decided that Georgia residents must be sacrificed for the whole, you have made that very clear. Please don't do the fee hike.[133]

Her testimony identified that the stipulation signed onto by the largest customers in Georgia reenacts a long-standing pattern of "low energy costs" for industry—seen as producing value in terms of economic growth—and "low wages" for residents who are devalued within the political economy of Georgia. Households must ultimately bear these energy costs as the burden of social reproduction is shifted onto them and already stressed charities and family networks. The deeply personal and yet widely shared experiences of stress over the light bill reported in public testimony, often by Black women, over months was disregarded by the Commission altogether.

Despite more than twenty thousand people speaking or writing in opposition,[134] the Commission voted a week later to increase the basic service fee for residential customers from $10 to $14 a month over three years, raising the average bill for customers using the smallest amount of electricity by just under $6 a month.[135] In total, a series of fee hikes, rate increases, and fuel adjustments means that electricity prices would rise more than $50 a month for the average Georgia resident by 2025. While not authorizing the full amount that the Company requested, the decision also set a 10.5 percent target return on equity. Chairman McDonald made a motion to limit this, citing the cost overruns at Vogtle did not deserve the Commission's reward, yet he could not get a second for his motion. Commissioner Tim Echols, who had missed the moving testimony of the public interest staff the prior week when he was in New York City for the S&P Global Platts awards ceremony, put forth the final rate plan, justifying that in New York, he had heard that the utility simply couldn't go below 10.5 percent and garner the capital it needed for its continued investments.[136] The capital structure and return granted to Georgia Power meant that it would not have to issue additional stock to cover its Vogtle construction costs.[137] An investment guide published in the week after the decision reported that the return was so generous, Southern Company stock should be seen "as an investment-grade bond with the state's PSC serving as a type of implied government insurance for its dividend."[138]

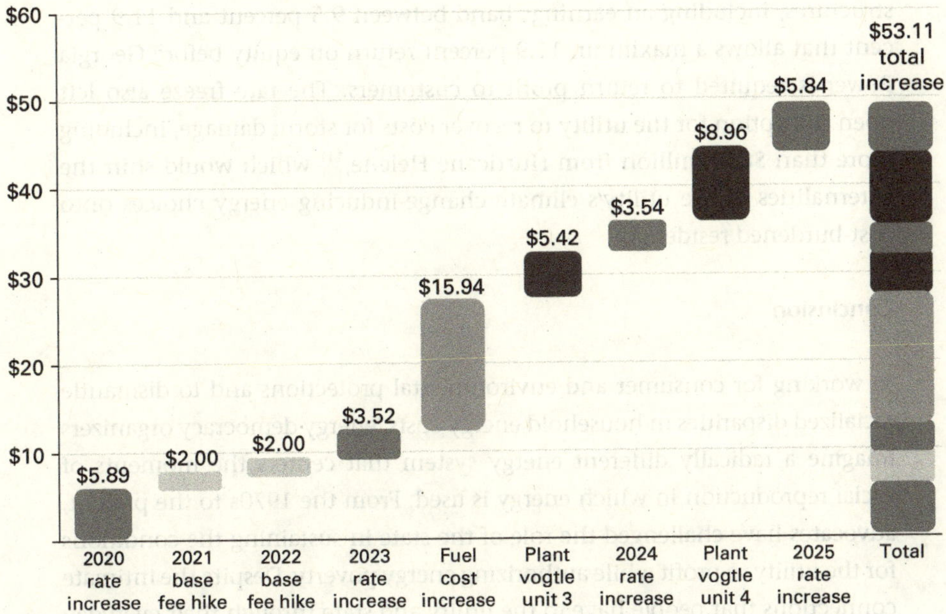

Figure 2.5
Expected monthly increases in average Georgia Power residential electric bills, 2020–2025. Graph by author, based on Southern Environmental Law Center analysis.[139]

From January 2019 to January 2020, the price of Southern Company stock increased from just under $45 a share to just over $70 a share. This dividend might be insured through the government yet is extracted from the electric bills of ratepayers.

In 2022, the Commission adopted another rate hike and in 2023 the final settlement for Vogtle was decided.[140] These decisions, in combination with gas price volatility following the 2022 Russian invasion of Ukraine, led residential rates to balloon (see figure 2.5). Electricity costs have drawn widespread condemnation, including failed legislative efforts to reinstate the Consumers' Utility Counsel and protect customers from costs associated with new data centers. Georgia Power projects data center load growth will lower costs for all customers by spreading the fixed costs for new generation to power facilities across a wider base of users. However, given widespread concerns, it agreed not to seek a scheduled rate increase in 2025. Despite freezing rates, advocates raised concerns that this decision eliminated an opportunity for the public to revisit the existing earnings structures, including an earnings band between 9.5 percent and 11.9 percent that allows a maximum 11.9 percent return on equity before Georgia Power is required to return profit to customers. The rate freeze also left open the option for the utility to recover costs for storm damage, including more than $800 million from Hurricane Helene,[141] which would shift the externalities of the utility's climate change-inducing energy choices onto cost-burdened residents.

Conclusion

In working for consumer and environmental protections and to dismantle racialized disparities in household energy costs, energy democracy organizers imagine a radically different energy system that centers the moments of social reproduction in which energy is used. From the 1970s to the present, advocates have challenged the role of the state in sustaining the conditions for the utility to profit while authorizing energy poverty. Despite the intimate connections that people have to the utility and state through their everyday uses of electricity, the Commission has disregarded the needs of poor people and people of color and refused to recognize how these moments of electric life create value for the utility and its shareholders and bondholders. Instead, it takes for granted the work of social reproduction that it has pushed onto households, charities, and houses of worship. Feminist urban political ecology draws into focus the labor of conserving power, applying for LIHEAP, and budgeting at the kitchen table to make the monthly electric bill that customers take on to pay utility rates and fees and manage the rate increases by the utility. As an industry that touches everyone in its captive service territory, the utility is a quasi-public institution and yet one empowered to provide services only to people able to pay. Acknowledging the unpaid labor of social reproduction that undergirds the grow-and-build model requires restructuring the calculation of utility finance and reasserting a role for the state to decommodify and provide universal service in the energy transition.

3 The Slow Violence of Fossil-Fueled Progress

Arching pines shaded the packed parking lot of the Lodge at Calloway Gardens where Southern Company shareholders gathered for the company's annual meeting. Golfcarts bustled back and forth, shuttling shareholders from their cars to the front doors. I had driven down from Atlanta with Sierra Club members, and we had arrived early to fill out paperwork to serve as proxies for Mercy Investments, a socially responsible investment fund that represents the financial interests of the Sisters of Mercy. The annual shareholder meeting has long been a site to lodge complaints regarding Southern Company business practices.[1] Beginning in the early 2000s, shareholder activists had pushed Southern to disclose their emissions and the impacts of aligning their business operations with efforts to limit global warming to 2 degrees Celsius. This campaign led Southern to commit to a no- to low-carbon portfolio for electricity generation by 2050.[2] Sierra Club volunteers had coordinated to use the open question-and-answer period hosted by then CEO Tom Fanning to ask about steps the Company was taking to realize its carbon neutral commitments. Atlanta residents were particularly concerned that Georgia Power's IRP, a twenty-year procurement strategy that the Public Service Commission reviews every three years, was not on track with its parent company commitments to achieve decarbonization and avert the worst effects of climate change locally or globally.

The meeting boasted of Southern Company's accomplishments and vision for the future of electric life. Throughout this book, I have used the term *electric life* to describe the functions, spaces, and times shaped by electricity, including both economic activity and our social lives. Displays integrated these quotidian and cutting-edge applications from a micro-grid technology company that Southern had recently acquired, a banner celebrating their recognition as one of the "50 best companies for diversity" by *Black Enterprise*, a hydroponics setup built by FFA members from the local high school, and a fleet of luxury electric vehicles. The crowd mingled over the exhibitions and a hearty breakfast buffet, which uniformed, Black resort staff kept stocked as they moved hot trays of bacon and carafes of coffee back and forth across the meeting, dodging the crowd. The room filled with local STEM students, diverse, mid-level managers at Southern affiliates, and less-polished shareholders, who were overwhelmingly elderly white people with the financial resources and time to travel to middle Georgia on a Wednesday. The spectacle omitted any discussion of customers in its companies' captive service territories whose bills finance accumulation for shareholders and the company.

At 10:00 a.m., the crowd filed into the darkened auditorium where the company board was already seated before an illuminated video screen. Tom Fanning started the event by giving a $10,000 donation to the Harris County High School STEM program. He thanked two workers who had helped to restore power after Hurricane Michael hit south Georgia without any reference to the climate change–fueled unpredictability of the slow-moving storm that rapidly intensified in the unusually warm waters of the Gulf of Mexico.[3] After a short presentation, he opened the floor to the audience and, for more than two hours, fielded questions ranging from the cost overruns at Plant Vogtle to emergent technologies and gas infrastructure. Fanning was charismatic, gracious, and deferential, embodying Georgia Power's slogan to be "a citizen wherever we serve." He gave complements and described shareholders who had come to the meeting before as dear friends. One shareholder, though critical of a gas compressor station in Valdosta, presented Fanning with homegrown squash, potatoes, and pickled okra.

Despite his smile and approachable demeanor, Fanning dodged questions from Sierra Club volunteers about the IRP underway at the Public Service Commission and the unequal burden of energy bills across racial lines. He explained that no Company talks more about the fact that "47% of their customers make less than $40,000," without mentioning any action to address the high cost of power for poor customers. When I asked whether Southern would support municipalities like Atlanta and Athens, which had committed to 100 percent clean energy in building power generation to meet these commitments, Fanning thanked me and smiled, saying, "We'll support everything that makes sense." He pushed back against the idea that any power source is no-carbon when you look at the supply chain and observed that it might be possible for one city to procure carbon-free electricity, but Southern Company as an entity operating above the fray of local politics had to be concerned with balancing electricity demands of the entire Southeast. Promoting the Company's nuclear investments, he assured me, in scornful reference to progressive organizing for a Green New Deal, that Southern offered a "green *real* deal. You can't find a better company to work with in advancing practical solutions for the future."[4]

His performance prophesied the Company's vision of enduring techno-modernist control over urban space and the future of electric life in which a solution to climate change will come, even if the utility is not yet ready to phase out coal or gas to implement it. Dipesh Chakrabarty observes that the anticipatory temporality of "not yet" operates within a universal historical logic in which progress is always on the horizon. Throughout history, this narrative has forced racialized, poor, and rural places and people—the under-developed—to wait for the conveniences of modern life.[5] The environmental consequences of racialized uneven development enact what Rob Nixon terms a "slow violence of delayed effects" that play out "not only over space, or bodies, or labor, or resources, but also over time."[6] Temporality is itself a relation of power that electricity contributes to shaping.[7] The utility upholds itself as a beacon of progress while downplaying or denying slow violence inflicted on people by polluting technologies that allow premature death.

The tactics of delay are calculated within the balance sheets of utility rates. In chapter 2, I focused on the rate-making process and the institutional and financial organization of electricity capital that contributes to racialized and classed disparities in energy affordability. Utilities finance power plant construction that is paid off over time through the monthly power bills of customers captive in their service territory. Once the utility makes an investment in a power plant, it must use it to recover its full investment. The process of paying off a plant can take decades. This means that coal burners have continued to operate despite the social and ecological costs of coal pollution. Given this financial incentive, the utility and its parent company have weighed in on legislation and regulation to shape the legal landscape governing pollution and to define both the pace and form of abatement and climate action. The practice of corporate lobbying to influence environmental law has been well established in social science and journalism.[8] Here,

I take this scholarship a step further to demonstrate how electricity capital relies on the life-times of people dispossessed of their health to serve the utility's modernizing vision.[9]

The utility's financial calculations undervalue people and environments in geographies exposed to pollution and routinize slow violence. Laura Pulido, Nicole Cotton, and Ellen Kohl observe that these choices force "devalued communities, places, and people [to] serve as pollution 'sinks.'" Communities rendered disposable bear the cost of the licit and illicit action by companies to evade implementing less-polluting technologies or cleaning up legacy waste and thus "enable firms to accumulate more surplus than would otherwise be possible."[10] Pollution from this view is not abstract but an intimate and on-going dispossession—a taking from the common share of clean air, land, water, and, thus, health—in daily life for frontline and fenceline communities to produce a different collective, yet privatized resource through the grid. This privatized infrastructure intersects with the privatized provision of social reproduction and externalizes the costs of asthma, chronic bronchitis, and cancer onto people exposed to the utility's waste. The time lost to sick days, emergency room visits, and lives cut short by preventable cancers or respiratory or pulmonary disease are not accounted for in the calculation of shareholder returns, yet this unrecognized biological and reproductive labor is nonetheless a subsidy to Southern's profit margin as the company asks captive ratepayers to wait for the cost of capital to be amortized before retiring coal plants or introducing renewables to the grid. So, too, the utility externalizes the costs of preparation and recovery from extreme weather fueled by climate change to which it contributes. Through the lens of feminist urban political ecology that centers these moments of social reproduction embedded in electricity infrastructures, I observe that the utility's vision of progress accepts the slow violence of coal, gas, and climate denial predicated on the devaluation of people and places to power electric life.

This chapter moves between the utility and its parent company to study the company narratives of progress that seek to control the pace and form of development and justify the fossil-fueled resource mix of Georgia Power despite the pernicious effects on human and environmental health. I begin with documenting the growth in energy intensity and the racialized narratives that undergirded the utility's drive to harness nature and labor for Southern industry as it transformed rural waterways and imported coal from across the country to fuel Atlanta's growth. As the costs of development became visible in smog clouding the city, the utility opposed environmental action, juxtaposing the unknowability of the effects of pollution and later climate change to what it asserted was the certainty of adverse economic impacts if mitigation measures were implemented. The utility contributed to fossil fuel industry campaigns that denied climate change and cultivated skepticism around climate science and action. As coal use has declined, the utility's management of legacy coal ash waste has relied on these same tactics to obscure environmental hazards for rural and urban residents. In the second half of the chapter, I consider the difficulty in building and sustaining environmental organizing to change resource practices at the utility by contrasting two coalitions. First, I study the challenges of building common cause around localized environmental effects in reviewing campaigns across the state to clean up legacy waste of coal ash. Second, I turn to successful coalition organizing by environmentalists and conservatives that united within the Green Tea Party to advance solar at the Commission. I conclude in asking about the work that remains to implement solar in line with the ambitions of energy democracy.

Although my focus in this chapter is on Georgia Power's efforts to escape the legal, regulatory, and scientifically grounded campaigns for environmental action, it is critical to recognize the intertwined history of environmental movements that preceded contemporary energy democracy organizing. Any change in the utility's resource procurement over time has resulted from hard-fought efforts of organizations like the Campaign for a Prosperous Georgia, the Sierra Club, Georgia Conservation Voters, and the Southern Environmental Law Center, who have corralled legal and financial resources to contest the utility in spaces such as the shareholder meeting. Notable critiques of the national environmental movement mourn its marginalization since the 1970s as advocates adopted technocratic and NIMBYist (not in my backyard) campaigns focused on leveraging scientific expertise to enforce environmental laws.[11] These tactics have proven limited in halting climate change or confronting the anti-intellectual climate denial of many in the contemporary Republican Party. The environmental movement has also been subject to critique by environmental justice scholars and organizers who argued that a focus on environmental protection failed to address the everyday environmental concerns that BIPOC and working-class communities live with.[12] Over time, environmental justice organizing shifted the environmental movement and its largely educated, wealthy, white, and Democratic base to advance a broader movement committed to justice and equity in the energy transition.[13] Across the decade covered in the second half of the chapter, I observe idiosyncratic, campaign-oriented coalitions between environmentalists, conservatives, or environmental justice communities that have proven difficult to sustain. These cases add to recent research on environmental populism that considers how social movements mobilize their lived experience as expertise to contest the ownership and impacts of energy infrastructures, as well as the challenges of sustaining alliances across lines of identity, class, and property ownership.[14] We can learn from these ephemeral alliances about the role environmentalism can play in broader democratic organizing. Nourishing these relationships to empower energy democracy requires asking how environmental protection and repair can integrate into a broader vision for a social wage through the infrastructures of electric life.

Resourcing Atlanta

Electricity was central to uneven processes of urbanization that directed coal, water, and mineral resources extracted from rural spaces to fuel Atlanta's growth.[15] The first coal-fired power plants built by Georgia Power's predecessors in downtown Atlanta burned coal to produce steam. By the early twentieth century, as these plants reached capacity, the utility began to look outside the city. Without abundant coal or gas resources available in other parts of the US, electric companies sought to harness the flow of waterways.[16] In his environmental history of southern rivers, Manganiello argues that hydropower development defies the canonical history of the South as a laggard region of the US: "The New South's men of capital and transnational engineers" recruited investors from New York, Toronto, and London to build a modern energy system that "consciously replaced free-flowing rivers with artificial reservoirs throughout the Southeast. They consolidated economic power in a decidedly unexceptional process and built a hydraulic waterscape, as their counterparts did in locales across the globe."[17] Reimagining the power of rivers as heretofore wasted resources, nature was put to work for a regional development agenda to emancipate "the average farmer's wife" from hard labor, power industry, and improve productivity as "man power is multiplied by machinery driven by electricity."[18] Even as Southern capitalists built the grid for the New South in reference to and in competition with other parts of the world, progress remained beholden to the memory of the Old South.

A commentary in 1924 described the potential of waterpower as exceeding the muscle power of enslaved people, comparing rivers and streams to "idle slaves [that] should be put to work."[19] In the decades following the abolition of slavery, nature was reimagined as a docile source of power that could reduce the need for Black workers to serve a white supremacist development agenda.[20]

Modernization was not without risk or opposition. One early effort to generate power from Tallulah Falls in Rabun County north of Atlanta faced protest from the Tallulah Falls Conservation Association, led by Mrs. Helen D. Longstreet, widow of Confederate general James Longstreet, who sought to preserve and "honor the traditions—and the traditional landscapes—of the Old South."[21] Steeped in the lore of the Lost Cause and outrage over Reconstruction, a fellow preservationist decried the involvement of a Northern company to "forcibly seize the loveliest jewel the state possesses."[22] Ultimately, the utility prevailed in court and finished the first of six dams it would build on the Tallulah and Tugaloo Rivers in 1913. Georgia Power would reach past the narratives of antebellum landscapes to legitimize the industrialization of Georgia's waterways, appropriating a Cherokee name given to a section of river for the powerhouse named "Terrora" to "intrigue the imagination."[23]

The damming of the Tallulah and Tugaloo Rivers created new natures described as even more marvelous than the rural landscape that had been flooded. A history of Atlanta describes, just as Atlantans "often stood in wonder and awe beside the tumbling torrent of Tallulah, the Terrible, there can be no doubt but that Tallulah shining in the streets of Atlanta, lighting her homes and turning the wheels of industry . . . is performing a service as beautiful as when her waters roared down the steep precipices of Tallulah Gorge."[24] The new natures of electric life were decidedly not "a natural resource, God-given," but rather, insisted company president Preston Arkwright, produced through the useful combination of labor and capital that brought sparkling wonder to the city.[25]

As uses of electricity grew, the utility faced growing pains of trying to match supply and demand. Many ambitious ventures for hydro development in north Georgia went bankrupt, realizing "that electricity is of no economic value in the absence of a market."[26] Lacking sufficient demand for current, Georgia Power's predecessor contracted the entire capacity of its hydro project to Duke's Southern Power Company. Promoting "renewable and cheap" energy "more easily manipulated than human labor," and not subject to the strikes and labor unrest that imperiled coal supply, Duke had lured papermills, lumber, and textile plants to its territory.[27] Companies shuttered New England factories to move to the Southern Piedmont where the share of national employment in textiles rose from 46 to 68 percent between 1923 and 1933.[28] During World War I, when coal was scare, the Georgia Railway and Power Company expanded its industrial contracts to supply hydropower to factories that had relied on in-house generators. A 1918 drought forced the company to curtail some of its industrial customers, and a subsequent drought in 1925 proved the danger of overreliance on the "variable flow" of rivers for the urban needs of electric life.[29] While the utility would go on to build dams on the Chattahoochee, Chattooga, Etowah, and Oconee Rivers and purchase hydropower stations on the Flint and Ocmulgee Rivers, it turned back to coal that did not eliminate but rather shifted dependence on rural landscapes that directed industrial, financial, and energy resources to Atlanta.[30]

Coal use in Georgia exploded with widespread consequences. By 1956, coal provided 75 percent of power demand with the utility constructing new power plants (see figure 3.1).[31] Georgia Power owned coalfields in Eastern Kentucky and Alabama and initiated long-term procurement contracts with nonunion Consolidation Coal in northeast Tennessee in the 1960s that delivered coal via a nonstop shuttle train from the mine mouth directly to Plant Branch.[32] Georgia Power's use of Appalachian coal drove ecologically destructive strip mining. In the 1970s, research from the Highlander Center estimated it would cost the average customer $0.50 a year if the utilities and mines discontinued the practice "but would save the land and lives of the people of Appalachia."[33] By the early 2000s, Georgia won the dubious distinction of purchasing the most coal mined through mountaintop removal.[34]

Coal also had pronounced local effects. In 1974, the Concerned Citizens of Monroe County organized to protest Plant Scherer, the last coal-fired power plant Georgia Power would complete, raising concerns about secrecy behind construction plans, use of eminent domain, and the possibility of coal ash ponds covering Indigenous mounds, as well as the flooding of timberland, hunting, and fishing grounds "literally teeming with bass, bream and catfish."[35] The plant was built over local opposition—in part because Georgia lacked a siting law to require a hearing before the utility could acquire land and the Commission did not question the need for additional capacity that the Plant would provide.[36] The utility touted the camping

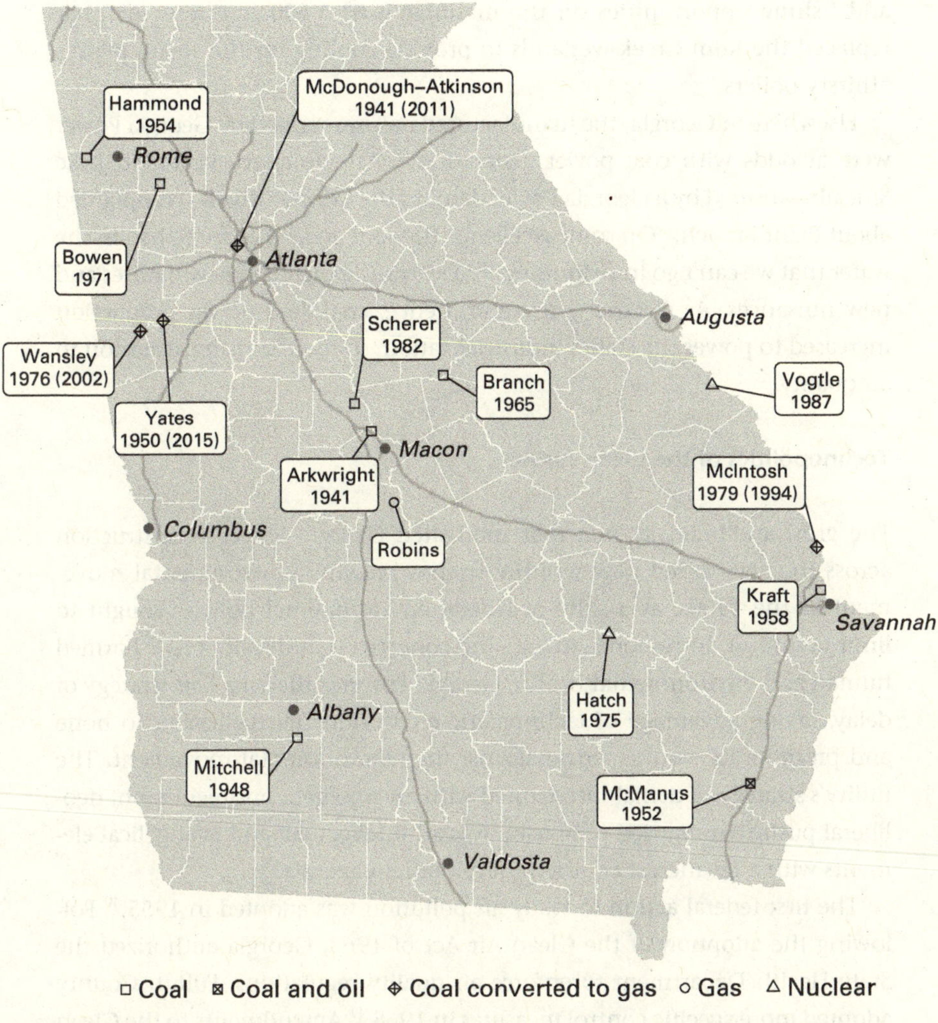

Figure 3.1

Map of Georgia Power coal, gas, and nuclear plants, including the date that plant first entered into service. Where applicable, the date a facility converted from coal to gas is recorded in parentheses. Map credit: Anna Mesa.

and fishing opportunities on the manufactured 3,600-acre reservoir that replaced the Rum Creek wetlands to provide cooling for the steam plant's "thirsty boilers."[37]

Elsewhere in Georgia, the manufactured natures created by Georgia Power were at odds with coal power generation.[38] As one angry visitor to Lake Sinclair—formed by a Georgia Power dam on the Oconee River—complained about Plant Branch: "On some weekends the soot and ashes are so bad in the water that we can't go in swimming."[39] Electric life included new natures and new nuisances. As electricity use and, hence, coal-fired power production increased to power the state's industrialization, so too did rising attention to air quality.

Technopolitics of the Clean Air Act

The grow-and-build agenda that motivated Georgia Power's construction across the state faced new scrutiny from a growing environmental movement in the 1960s, as a suite of federal environmental policies sought to limit pollution. In response to growing concerns that development harmed human and environmental health, Georgia Power initiated a licit strategy of delay, taking advantage of technocratic environmental regulation to hone and promote its own expertise against the environmental movement. The utility's strategy of denial intertwined with an emergent antiregulation, neoliberal push that married pro-business, anti-intellectual, and evangelical elements with a revanchist Southern development agenda.

The first federal action to study air pollution was adopted in 1955.[40] Following the adoption of the Clean Air Act of 1963, Georgia authorized the State Health Department to enforce air quality regulations. Fulton County adopted more specific control measures in 1968.[41] Amendments to the Clean Air Act in 1970 substantially expanded federal authority to limit emissions from industrial and mobile sources of pollution. This included establishing National Ambient Air Quality Standards to regulate criteria air pollutants, sulfur dioxide, particulate matter, carbon monoxide, nitrogen dioxide, lead, and ozone, which states were expected to monitor under State Implementation Plans reported to the newly created Environmental Protection Agency (EPA).[42]

Monitoring systems placed Atlanta in the bottom third of cities nationally for air quality. Transportation was and remains the most significant source of pollution, but Georgia Power's facilities were primary stationary sources of concern.[43] Facing regulation at the local, state, and federal levels, Southern Company chairman Harlee Branch Jr. disparaged "today's noisy 'environmental evangelists'" in an address to the Atlanta Chamber of Commerce in 1970, stating that "pollution abatement has been a part of the tradition in power generation for years."[44] Setting an example that Tom Fanning would follow half a century later, Branch cautioned against environmentalist hysteria and urged trust in the Southern Company's homegrown approach to conservation.[45] As Georgia Power promised it was taking necessary steps to address pollution at facilities of concern near Atlanta, local environmental organizing turned to focus on water pollution and road construction.[46] The complex jargon that shrouded State Implementation Plans further contributed to the professionalization of environmental regulation enforced between nonprofits, state agencies, and corporations obligated to adhere to the new framework of federal environmental law. Sellers notes that "corporations like Georgia Power gained a growing ability to dominate [environmental enforcement] sheerly by dint of [employing] all the expertise they could afford."[47]

When the EPA proposed additional rules to prevent deterioration of air quality in areas that were in attainment under the Clean Air Act, which would have affected Georgia, Georgia Power used their influence and expertise to lobby against tightening regulation that threatened the growth agenda of the state and urban regime. Company President Scherer wrote to Presidents Ford and then Carter, arguing that proposed emission controls scapegoated big business and risked exacerbating unemployment, inflation, and rate hikes for "questionable health results."[48] Georgia Governor Busbee weighed in to assert that more stringent federal regulation would "severely" restrict "industrial development in Georgia and elsewhere in the South."[49]

Environmentalists, unions, and public officials in the Northeast and Midwest championed the EPA for attempting to curtail capital flight to "blue sky regions" in the South and West to avoid strict abatement measures that had already been enacted in more polluted regions. The editorial board of the *New York Times* endorsed efforts to strengthen the Clean Air Act (alongside proposed revisions to labor law that ultimately did not pass), observing these measures were "designed to serve national needs, not narrow regional interest. But in the process, the maturing economy of the South could be forced to pay fairer dues to its own workers and to the nation."[50] Strengthened environmental laws intended to curtail the race to the bottom, curb the devaluation of labor and human and environmental life, and protect air and water quality for the benefit of all. The Clean Air Act amendments that were ultimately enacted in 1977 introduced the "New Source Review" rule, requiring new stationary sources to adopt best pollution control technologies for SO_x, NO_x, and particulates. This effectively mandated use of scrubbers for coal-fired power plants. However, generators built before 1977 that did not undergo modifications beyond routine maintenance were exempt based on the assumption that antiquated facilities would soon be retired.[51]

Rather than close aging facilities or introduce pollution control technologies, Georgia Power and its parent company practiced strategic evasion. First, Southern looked for low-sulfur coal, including from apartheid South Africa. When first proposed in 1975, Southern was forced to backtrack after facing opposition from the Georgia Power Project, civil rights groups, and public officials. Members of the Georgia General Assembly's Black Caucus and Atlanta City Councilors charged the deal was for "the world's most expensive coal when measured in terms of human dignity and workers' rights."[52] While stopped at Georgia Power, in 1976, Southern Company signed a ten-year purchase agreement with Transvaal Coal Owners Association to supply Gulf Power over objections from stockholders, touting a guarantee that enslaved and convict labor was prohibited in mines.[53] Southern continued to promote the labor and human rights records of its South African coal suppliers as a reason for its continued import of coal despite objections and repeated shareholder resolutions in opposition.[54] Southern also looked to nonunion coalfields in the Powder River Basin of Wyoming that had a lower sulfur content than Appalachian coal.[55] Lower-cost coal was seen as a way "to provide even more energy value" if at the cost of weakened labor standards.[56]

Alongside the shift in coal procurement, Southern Company skirted compliance by retrofitting facilities to produce more electricity without upgrading pollution controls. In 1999, EPA and the Department of Justice sued nine utility companies, including Georgia Power, alleging that "a large number of coal plants may have profited by polluting the air beyond the legal health-based limits of the law."[57] Georgia Power's Plants Bowen and Scherer were named in the lawsuit and Plants Branch, Hammond, McDonough, Wansley, and Yates were also targeted for stricter abatement measures.[58]

Southern Company pushed back against the suit and what it regarded as an unfair change in the Clean Air Act standards. Their "perfectly legal and widely practiced" investment in lobbyists and campaign contributions paid dividends during the George W. Bush administration as the New Source rules were taken under review. The company also used the ubiquity of electric life to their advantage. Georgia Representative Bob Barr admitted that Southern Company lobbyists were convincing in "mak[ing] the claim that every single constituent of that [house] member or senator is touched by their industry or business."[59] When questioned about lobbying, one company official explained, "We feel our obligation is to be there, telling members of Congress, especially in the states we serve, what is the cost of all of this and what are the consequences."[60] The utility downplayed its culpability and played up the broader economic costs of environmental policy—or at least those measurable on its balance sheet—and in doing so laid the groundwork for a strategy of weaponizing its expertise to cultivate environmental skepticism, which it would continue to exploit as concern about climate change mounted.

Southern's accounting of the consequences downplayed the costs to environmental and human health even after metro Atlanta repeatedly failed to reach attainment of acceptable levels of criteria air pollutants after ground-level ozone standards were set and tightened by the EPA in the 1990s.[61] Local environmental organizing led by the Campaign for a Prosperous Georgia placed a significant share of the blame on Georgia Power. By 1998, Southern Company had become the second-worst emitter of CO_2 and SO_2 and likely among the top three emitters of mercury in the US.[62] Southern purchased allowances to cover SO_2 emissions for two-thirds of its plants. At Plant Bowen, which had the highest cap of allowable emissions of any facility nationwide, SO_2 emissions remained 40 percent above EPA targets.[63] Southern also ranked as the third-worst single-source emitter in the country of NO_x, which interacts with sunlight to form ground-level ozone, which is especially dangerous to people with respiratory diseases and children who are more likely to develop asthma.[64]

Throughout the 1990s, accumulating evidence demonstrated the link between ozone pollution and childhood asthma, with distinct racialized disparities between Black, Latino, and white children, produced through the racialized uneven development of housing, transportation, healthcare, and food that magnify "discriminatory geographies of pollution."[65] The willful disregard of the power company added to the slow violence of the built environment.

The "asthma upsurge" in Atlanta was widely reported on locally. In an interview with the *Atlanta Journal-Constitution*, an eleven-year-old described not being able to play outside because "I start wheezing a lot and it gets hard to catch my breath, and I have to go inside." His mom reported, "I really pay close attention for the announcements about smog alert days," when she had to keep her son inside, even when "He will beg me to let him go out, and it breaks my heart that he can't."[66] Another woman weighed the decision to move from Atlanta after being hospitalized for nine days following an asthma attack, saying, "I love Atlanta. I was born here, but I hate to see it like this where you have trouble breathing."[67] Georgia Power's strategies to evade pollution regulation externalized the cost of power production onto Atlanta residents who embodied their smoggy environment. Individuals and families were forced to take on the social reproductive labor and financial costs of healthcare. A report aptly titled "Power to Kill" published by the Clean Air Task Force estimated that emissions exceeding the Clean Air Act caused between 289 and 691 preventable deaths in Georgia and between 5,237 and 11,841 preventable asthma attacks in the 1990s. Wind flow patterns blew toxins from Atlanta north in "a transport of death and disease" spreading particulate matter also linked to acid rain that degrades soils and harms plants.[68]

Air pollution united environmental and antipoverty and racial justice organizing in Atlanta as an issue of environmental justice, driving environmental and civil rights organizations to work more closely together. Despite growing public attention from a diverse coalition, Georgia's representatives at the state and federal level failed to act against the utility, bowing to Georgia Power's lobbying efforts, and instead began to espouse an antienvironmental position, cemented as political winds in Georgia shifted from majority representation by the "ideologically diverse" Democratic Party to Republican control.[69] This shift would propel attacks on the legal and scientific foundation of "environmental regulatory structures and practices" and also undermine new federal actions on environmental justice, which advocates had worked to achieve at the EPA.[70]

Ultimately, the Supreme Court rejected the Bush administration's industry-friendly interpretation of the New Source Review provision.[71] Despite winning the sympathy of elected officials, Georgia Power was required to install pollution-abatement technologies at its facilities in 2006, nearly thirty years after the 1977 New Source Rule was promulgated. Plants Wansley, Bowen, and Hammond were prioritized, given the immediate impacts on Atlanta's air quality.[72] Plant Scherer also received scrubbers, and other plants were retired or converted to gas.[73]

Pollution abatement worked to mitigate health impacts, if not addressing climate concerns. A 2023 study linked coal plants in Georgia to tens of deaths annually in contrast to the thousands attributed to unabated pollution in the early 2000s.[74] In 2022, Georgia Power announced that because coal plants are "less economically viable," it would retire and decertify all remaining coal burners by 2028, except Plant Bowen Units 1 and 2, which it plans to keep online through 2035.[75] However, with rising power demand due to data centers and large loads, Georgia Power reversed course, filling for permission to continue operating Plants Bowen and Scherer into the 2030s. The utility has also increasingly relied on methane gas, receiving Commission approval in 2024 to build 1.4 gigawatts of new methane gas generation and requesting permission in 2025 to certify construction and procurement totaling nearly 10 gigawatts, including close to 6 gigawatts of gas.[76] Methane emits significantly less NO_x and PM2.5 but is a more potent greenhouse gas. As coal use across the state changes, the question of how to manage its legacy remains.

Coal Ash Cleanup and the Intimacies of the Power Bill

While Georgia Power's resource mix has shifted from coal to gas, pollution from coal waste remains in twenty-nine coal ash ponds that cover 2,300 acres adjacent to its eleven coal plants across the state.[77] These expansive lagoons store millions of tons of coal ash, the dusty residual by-products of burning coal that are mixed with water to form a slurry that contains heavy metals and radioactive materials including arsenic, selenium, mercury, boron, cobalt, radium, sulphate, nickel, and lead.[78] Coal ash is the second largest waste stream in the US, behind municipal solid waste. A long track record of spills accompanies Georgia Power's air pollution emissions violations, which have presented an on-going concern for the environmental movement.[79]

Coal ash began to be taken more seriously after a catastrophic spill at the TVA Kingston Plant in 2008.[80] Following a subsequent spill in North Carolina in 2014, the EPA initiated a rule-making process to limit dispersion of coal ash dust, prevent slurry leaks and groundwater contamination, mitigate the chance of dam failures, and set rules for "beneficial" reuse of coal ash in building materials, such as Portland cement and drywall.[81] In response, Georgia Power announced a $2 billion plan to retire its ash ponds. New ash produced from on-going coal production is handled dry, and nearly all of it is sold to manufacture building materials.[82] Yet, the processes for closing existing ponds remain contested. Advocates pushed the utility to follow the example of Duke Power, which was required by the state of North Carolina to remove coal ash from unlined pits and dispose of it in lined landfills after floodwaters from Hurricane Florence breached several ponds.[83]

Instead, Georgia has taken a different approach, negotiated through the Commission, the General Assembly, and the EPA. In 2016, Congress adopted the Water Infrastructure Improvements for the Nation (WIIN) Act, which gave states the authority to operate coal ash permit programs if the EPA determines those standards are as protective as federal law.[84] In 2018, the Georgia Environmental Protection Division (EPD) submitted its application to initiate a permitting process. Given the utility's influence and "track record of delaying or blocking efforts to regulate pollution,"[85] advocates raised procedural and site-specific concerns that the EPD would greenlight Georgia Power's request to cap ten of their unlined coal ash ponds in place. A letter cosigned by the Southern Environmental Law Center, Altamaha Riverkeeper, Chattahoochee Riverkeeper, and Coosa River Basin Initiative pointed to Georgia Power's own disclosure that its coal ash ponds were leaking and "contaminating groundwater with one or more toxic pollutants at levels exceeding the groundwater protection standard or other health-based standards."[86] Despite significant opposition at an EPA hearing in Atlanta, the Trump administration EPA approved Georgia's request.[87]

Opposition to the devolution of regulatory authority to cap wells in place focused on two interrelated issues: the cost of cleanup and on-going risk of pollution from coal ash submerged in groundwater.[88] In 2019, the Public Service Commission allowed Georgia Power to recover a first tranche of $525 million for coal ash cleanup.[89] In a town hall in Athens, Commissioner Echols explained that to not allow Georgia Power to recover the costs would not be "fair" as the utility had been doing everything "right" until EPA rules changed.[90] The idea that Georgia Power is doing everything within the law ignores the industry's role in constructing the licit practices of waste handling.[91] The Sierra Club sued to halt the rate increase, arguing that the Commission "failed to take into account Georgia Power's culpability in creating the coal ash problem to begin with, and thus should not be allowed to pass all of those costs onto customers."[92] Simultaneously, the Sierra Club formed a statewide coalition of environmental and environmental justice organizations to weigh in on permitting processes for the ash pond closures.

In 2021, the state Environmental Protection Division convened its first public hearing on Zoom to hear feedback on a proposed permit to dewater a coal ash lagoon adjacent to Plant Hammond and treat the runoff through a complex discharge system that would eventually be released into the Coosa River. Afterward, it would cover the pile of dust in concrete. Over the course of the two-and-a-half hour meeting that 130 people from across the state attended, only Aaron Mitchell, Georgia Power's director of environmental affairs, spoke in favor of the plan to cap the waste in place.[93] Attendees raised a host of concerns about the social reproductive effects of coal ash. Several women testified to reproductive health problems and cancer clusters in their towns, linking intergenerational health impacts to choices by the utility.[94] Vicki bluntly added that removal was necessary, as "Coal ash will leak into the ground water. It will migrate into the creeks and the rivers where the toxins will make people sick. It will kill people." We know that "Georgia is number one for business," explained Treva, but "We're asking Georgia to take care of their people this time and not put profits over our people."[95] Many joined her in condemning the utility for adopting cap-in-place as the cheapest method of disposal, even as the costs had risen from Georgia Power's initial $2 billion estimate in 2016 to a staggering $9 billion price tag.[96]

The cap-in-place disposal strategy, which speakers condemned as "a plan to let communities be poisoned,"[97] is a new chapter of the utility's process of stripping resources from rural environments. Resources include not only coal and hydropower but environmental quality and health.[98] The human toll of Georgia Power's disregard for its neighbors has been most loudly protested by residents of Juliette in Monroe County, who live in the shadow of Plant Scherer, and for years preceding the new regulation of coal ash had documented nosebleeds, neurological problems, and rare cancers as the ash impoundment adjacent to the plant filled.[99] When Scherer opened in 1982, the Company downplayed documented health risks linked to coal to avoid investing $95 million in a liner for the slurry pond—equivalent to $300 million in 2023 dollars, or a fraction of the costs it will now take to clean up the waste. After the EPA required Georgia Power to monitor groundwater for leaks, Georgia Power began to purchase homes adjacent to and downstream from Scherer and four other plants, spending $15 million on 1,900 acres of land.[100] While ostensibly to prevent inconvenience as it begins to close its coal ash ponds, residents who sell are bound by nondisclosure agreements.

Former homeowners and their families have hinted that the process also allows Georgia Power to cap wells and prevent water testing that might reveal their culpability for groundwater contamination and necessitate full excavation.[101]

The exposure to slow violence in Juliette and other peri-urban areas in Georgia is intertwined with and yet geographically distinct from air pollution that concentrated in the Atlanta metro area. Juliette complicates the story of Georgia Power's practice of environmental racism as the most vocal critics of the coal ash impoundments are white. The production of white rural space and vulnerability to environmental racism in Georgia are in themselves linked to electrification and a century-long process of urbanization that shifted rural development. Rural Georgia became whiter over the twentieth century as a product of racial terror and freedom dreams of Black people leaving slavery, sharecropping, and racist bankers and extension agents who denied Black farmers access to credit and resources to stay on the land, including financing for electrified irrigation, pumps, and feed grinders and diesel-powered heavy equipment that increased farm productivity.[102] Black farmworkers and farm owners migrated north, west, and into cities where Marable argues they were "coerced to work at low wages while paying skyrocketing utility bills." As a result, he continues "Getting 'power to the people' has become just another way of maintaining and expanding the system of racial and economic suppression in the New South."[103] Today, Monroe County has just shy of thirty thousand residents, about three-quarters of whom identify as white. Since Plant Scherer came online, the county has grown slightly but also become less racially diverse as the Black population has remained relatively constant. About 21 percent of residents identify as Black compared to 38 percent in 1980.[104]

In her ethnographic study of Burke County where Vogtle is located, Loka Ashwood observes what she terms a "utilitarian" drive to the utility's growth imperative based in racialized logics rooted in settler colonialism and slavery that renders people, their health, and familial and social attachments to place secondary to development. She asserts this utilitarian tendency is "growing in unanticipated ways. Profit now subjects to utilitarian sacrifice those [landowners, especially white landowners] once immune to it."[105] The "utilitarian" least-cost assessment assumes that rural people who were promised jobs and more tax revenue than all the other properties in the county combined as a result of new electricity generation infrastructure would accept risks that come with power plants, even as land, water, and life-time are extracted to fuel growth elsewhere.

Extraction and dispossession from "forgotten places" is invisible in most people's quotidian uses of electricity. Yet, Gilmore writes, the feeling of living in a forgotten place can engender solidarity against licit extraction shaped through white supremacy that accepts vulnerability to premature death. She observes: "In the United States today, *white* people suffering from a concentration of environmental harms in some rural communities have learned to call what is happening 'environmental racism' without imagining that they are somehow excluding themselves from the analysis and instead feeling whiteness peel away in the context of their vulnerability. This stretched understanding of racism enables vulnerable people to consider the ways in which harmful forces might be disciplined and harms remedied."[106] Gilmore asks whether and how experiences of slow, structural, and state violence can reorganize racial identification and enable white people to articulate a politics of solidarity with people of color rendered disposable to the process of accumulation. Instead of racial identity, she asks how can "broadly contested sensibilities—indeed *feelings*—[be] the basis for political struggle?"[107] Shared grievance against the utility could be a fertile ground for coalition building if the injuries that differentiate and connect urban and rural space can be linked.

Yet, politics are contextual and unfold in time and space. In the 1970s, Monroe County residents pushed back against Georgia Power attempting to redbait Atlanta organizers who had come to support the Concerned Citizens of Monroe County in their fight against Plant Scherer before construction began, arguing that the utility attempting to coerce land purchases was of more significant concern to the community than socialism.[108] Today, building connections between Atlanta-based Sierra Club members working to stop Georgia Power's lax coal ash containment plans and Juliette residents fighting for redress against Plant Scherer is mired in misrecognition. In 2018, as Georgia was seeking to get control over coal ash regulation, the Sierra Club Beyond Coal campaign engaged in the coal ash fight with the goal to knit together a statewide coalition. Michael, a white environmental organizer, described that "part of what we were thinking was build outrage around coal ash, because that is an imminent issue."[109] Volunteers strived to drive participation in on-going regulatory proceedings and build solidarity between fenceline residents alarmed by leaky coal ash ponds or poor air quality, coastal and south

Georgians affected by climate-induced natural disasters, and organizations in Atlanta that were already mobilized around the cost of electricity. In other words, recognizing both the possibilities and pitfalls of militant particularisms that can cripple environmental justice struggles, the Sierra Club highlighted the connections between constituencies organizing to change the utility and state practices that allowed slow violence to continue.[110]

In the years of the first Trump administration, COVID, and the Black Lives Matter movement, recognition of a shared enemy was easier for environmentalists to see than the communities they attempted to work alongside. On one hand, COVID made digital organizing more possible, bringing together powerful statewide opposition as seen in the hearing against the Hammond permit. On the other hand, the "feelings" that Gilmore describes didn't always connect. The Sierra Club had some success in the "bluer" counties where they plugged into local colleges, yet, Michael continued, "We stayed away from organizing around Scherer, at the request of people who were already organizing there," because the "Trump red-hat wearing, solid rural conservatives all white, basically" viewed Atlantans as agitating outsiders who could polarize or potentially jeopardize local organizing.[111] Juliette residents leaned into their Republican affiliation to lobby legislators and the governor, distancing themselves from an environmentalist left, to push legislation requiring excavation of the ash ponds.[112] Even their lobbying has failed to move the General Assembly to address coal ash. In 2024, Georgia Power settled a lawsuit out of court with Juliette residents for an undisclosed sum.[113]

Organizing around coal ash also proved challenging as intervention points to drive participation and signal opposition to the utility's plan crossed the Public Service Commission, which had to approve costs, and the state EPD and federal EPA, which would review the adequacy of cleanup plans. A Georgia Power neighbor facing poison in their drinking water and customer subsidizing the cost of this multibillion-dollar cleanup project were focused on different regulatory processes. Michael explained: "Everybody pays a power bill so isn't that a natural constituency to turn against the greedy evil power company? No, it's not that easy."[114] In 2022, the Georgia Supreme Court ruled against the Sierra Club and allowed the Commission's decision to pass through the costs of coal ash cleanup to customers to stand. This decision was based on the premise that Georgia Power's cleanup strategy was acceptable to the EPD. In turn, the EPD Land Branch Chief told Juliette residents in 2020 that the utility would satisfy their standards for proper disposal if it could "show its contaminants had not migrated beyond its property lines,"[115] hence, Georgia Power's land purchases. Land purchases pale in comparison to the $3 billion estimated additional price tag if the utility were required to excavate ash and dispose of it in lined landfills. This additional expense, combined with the charge of past negligence, could void the utility's attempt to recover the cost of cleanup. Details on the technology and calculation of these estimated costs have been redacted from Commission files as a trade secret, which the utility asserts "derives economic value from not generally being known." These redacted details contain incalculable life-times. The Commission and EPD have accepted costs to human health, and the utility's plausible deniability and assertion of unknowability of harms, in allowing cleanup costs to be added to the rate base and to sustain shareholder value.

Learning to "Speak Republican" and Solar Advocacy at the Commission

The utility's disregard for the externalities of power production on people and the environment reaches its apex in its studied disavowal of climate change. The refusal to curtail planet-warming carbon, methane, and other greenhouse gas emissions represents its ultimate effort of delay in the interest of maintaining electricity capital. Where misrecognition stymied coalition efforts to organize against coal pollution, conservative and environmental organizations built common cause to advance solar before the Commission in proceedings around the utility's IRPs.

In the 1950s, Georgia Power embarked on a capacious and exploratory research agenda, including early participation in nuclear and solar studies. Through the 1970s, utility officials argued solar cells could utilize locally abundant energy from the sun and even lobbied for state tax credits to keep money within Georgia instead of spending to import oil and coal (a point that solar energy advocates still make).[116] The utility built a solar demonstration project, put solar on the roof of its headquarters, and built Future I, a model solar electric home, in downtown Atlanta as part of a planned residential community. It even sponsored a pilot project to use electric buses in Fulton County during the Atlanta Olympics.[117]

Despite this promising early investment, beginning in the 1990s, as the cost of pollution compliance mounted, Southern Company pivoted to work with the Edison Electric Institute to spearhead a campaign to "reposition global warming as theory (not fact)."[118] Undercutting early consensus that guided the Clean Air Act's adoption that "people cause pollution,"[119] Georgia Power's parent company honed climate skepticism. Southern Company partnered with Exxon and Chevron to propose recruiting climate skeptic scientists to lodge public opposition to the Kyoto Protocol. While the project was never formally launched amid public outcry, the mobilization of utility expertise, research capacity, and political power to discredit climate science continues in less overt ways.

Southern Company still avoids climate change. Climate is not mentioned in the 2023 annual report and appears in their 2024 report only in reference to the second Trump administration's withdraw from the Paris Agreement.[120] Their filing to the Securities and Exchange Commission in 2022 conflated climate change and climate activism as equal risks.[121] One advocate explained, "Georgia Power says they refuse to talk about it, because the Commission doesn't believe it's real, so we don't talk to George Power about it, outwardly."[122] The refusal to take on the present imperative of climate change is a manifestation of the progressive politics of time that is built to sustain the status quo, allowing the utility to control the direction of change or delay action until it might benefit them.[123] In doing so, the utility, with the approval of the Commission, has gambled with the life-times of people forced to bear the privatized cost of its generation choices from asthma medication to extreme weather recovery.

Given the refusal to see climate change as a present concern, the benefits of renewable energies have been proven and won politically on other terms. In 2013, an unconventional alliance formed to push the Commission and Georgia Power to invest in solar. To that point, the utility had only built exploratory or demonstration projects. During the IRP, the Sierra Club, Republican Commissioner Bubba McDonald, and Atlanta Tea Party founder Debbie Dooley championed solar, leading the Commission to require Georgia Power to add 525 megawatts of solar. Commissioners still couched the decision in free market language: "This motion is not a mandate," insisted Commissioner McDonald in a press release.[124]

Following the Commission's decision, Sierra Club employees (not with the formal endorsement of the organization), solar industry representatives, and the Atlanta Tea Party joined to form the "Green Tea Coalition" to advance shared priorities. Campaigns included working against a fee for residential solar installation in the 2013 rate case that followed the IRP and supporting a legislative effort to allow third-party financing for solar installation through the Solar Power Free-Market Financing Act of 2015.[125] Political solidarity was based not around climate but cost.[126] Commissioner Echols, who has gone on to champion solar, storage, and EVs, described the demonstrated support for solar across the political spectrum as a "game changer."[127] He has urged others to demonstrate how clean energy fits into a Republican playbook, saying, "Don't come into my office talking about climate change or the environment. . . . Talk about new jobs, talk about low-cost energy, talk about reduction of transmission lines. . . . Learn to speak Republican here."[128] Despite its impact, the Green Tea Coalition stalled—its champions at the Sierra Club moved on and the chair of the Atlanta Tea Party began organizing regionally with a new group, Conservatives for Energy Freedom.[129] Other conservative organizations have made appearances at the Commission, including the Evangelical Environmental Network, for whom implementing renewable energy to mitigate climate change is about a moral responsibility to the unborn.[130] The appeal to a Christian ethos of care attempts to leverage the Christian religiosity of the Commissioners, who often describe the intersections of their faith and orientation to public service, to push Georgia Power.[131] Notably, the Commission has maintained solar in subsequent IRPs as an investment in Georgia's growing clean energy economy, even as national survey data indicates that conservative support for clean energy as a political priority for voters has plummeted since 2018.[132]

The loudest voices for clean energy before the Commission are now the solar industry, environmental, energy justice, faith, and consumer advocates. In the 2019 IRP and rate case, these organizations came together to support net-metering for solar. Net-metering is a financial incentive for utilities to buy back power generated through residential solar that is not used at the source and returned to the grid. The program has been among the most effective policies in expanding residential solar adoption, yet only those with rooftop solar can participate.[133] The Commission required the utility to pilot a version of net-metering for five thousand homes,[134] which was oversubscribed by 2021 and has not reopened as of September 2025 despite overwhelmingly support for the program in public comment at the Commission. Commissioners have cited the regressive possibility of cross-subsidization, meaning customers without solar are obligated to pay for the program for households with solar who benefit from net-metering.[135] Alternative programs to expand access to clean energy like community solar have also been stymied.[136] As part of the same stipulation, Georgia Power also agreed to establish income-qualified, residential energy efficiency programs, including a pilot for on-bill financing of energy efficiency, which advocates supported as a way to democratize the benefits of clean energy.[137] Without income-qualified programs, energy upgrades are available only to homeowners with the upfront capital to spend implementing these technologies on their property. For instance, a survey of solar installations in Atlanta found most are in wealthy, whiter neighborhoods, or on recently purchased homes in gentrifying neighborhoods.[138] Similar disparities have been documented in other utilities in the use of utility rebate programs for energy efficiency.[139]

In the 2022 rate case, the Commission approved another Distributed Energy Resource pilot proposed by Georgia Power through which the utility will operate distributed solar and battery energy storage systems installed at commercial and industrial sites. This program reflects a willingness to revise the unidirectional flow of electricity from the power plant to sites of consumption, while the utility still provides central oversight and planning of generation and storage assets. Importantly, for the utility, it is then able to recover the costs of distributed energy resources within its regulatory model. A Georgia Power official described that "limiting DER to those maintained and operated by Georgia Power ensures an appropriate level of operational oversight, decreases the operational risk profile of the program overall, and provides greater administrative efficiency," while still realizing "demand response value."[140] The pilot program is estimated to produce 250 megawatts of capacity—as much as a gas plant.[141] Management by the utility means that Georgia Power can recruit solar and storage from areas where the grid is most congested, minimizing the need for new transmission and distribution infrastructure, which are among the steepest costs driving up the price of electricity.[142] The utility is also able to promote this new green commodity to commercial and industrial customers coming to Georgia. These programs offer additional revenue to property owners who can benefit from bill credits from their participation. In its 2025 IRP, the utility requested permission to procure an additional 100 megawatts of distributed energy resources, including from residential customers, and described its efforts to deploy a Distributed Energy Resource Management System.[143]

As the utility begins to recognize "demand response value," offering property owners who host these assets the chance to see lower bills or even revenue, propertyless and poor customers are kept waiting. This partial response to demands raised by advocates to expand renewable energy and reliability falls short in addressing on-going concerns about affordability and democratic pathways for all to participate in distributed energy systems. If implemented in combination with the pilot for on-bill energy efficiency, the expansion of the utility's distributed energy resource procurement could be an exciting potential pathway to enable the utility to bring distributed generation into its business model and to democratize access by subsidizing expansion of residential solar and storage as a virtual power plant that provides resilience and lowers bills.[144]

Despite these promising pilots, the utility has moved at a glacial pace on implementation in contrast to its mass expansion of gas to meet demand.[145] As it sets the pace of progress and dictates the value of everyday and generational times imbricated in its business of selling power, the company's willingness to sacrifice future comfort, safety, and continued vulnerability to the slow violence of climate change remains core to its model of making a profit from the daily needs for electricity.

Conclusion

The politics of the power plant require studying the long arc of the socioecological transformation of Georgia through urbanization to understand the combination of human energies, environmental capacity to absorb pollution, and energy resources that sustain the utility. The resource decisions of the utility and Commission render people in Georgia disposable, exposing downwind and downstream residents to air pollution and groundwater contamination. The utility has simultaneously and unevenly devalued its neighbors while creating new opportunities for accumulation as it begins to recognize the value of distributed energy, energy efficiency, and dispatchable storage through demand-side management programs. Suturing the questions of groundwater contamination from coal ash, climate change, and racialized inequity in utility bills and access to clean energy technologies remains an on-going challenge to build popular support for energy democracy in Atlanta and across Georgia. Tracing the time of social reproduction through feminist urban political ecology is one starting point to study the life-times enmeshed in the changing geographies of the power plant. The question of who bears the costs of pollution and benefits as power production changes has been widely discussed in academic and policy literature as issues of procedural and distributional justice.[146] Yet this history also demonstrates the challenge of maintaining solidarity organizing to move beyond a liberal or harm reduction framework and empower democratic governance of the energy system to achieve energy justice in which the utility "stop[s] poisoning people."[147] As Katherine McKittrick asserts, to confront the calculative logic of the utility that exploits the intertwined relations of "race, space, and premature and preventable death" demands new knowledge systems rooted in "human life."[148] It demands making room for livingness and embracing the temporalities of electric life within and outside of the utility's control as moments of connection to forge solidarity. To begin to dismantle infrastructures of slow violence requires refusing the utility's impoverished, linear narrative of progress toward environmental quality on a future horizon and its current practices of rendering people disposable to build instead democratic governance for a public utility invested in social and environmental health.[149]

4 Organizing Electricity on and off the Job

In September 2023, the US Department of Energy hosted a workshop in Atlanta that brought environmentalists, contractors, and public officials to the IBEW Local 613 union hall. The aim was to support community-labor union collaboration to advance worker organizing and racial equity in new clean energy projects funded through BIL and IRA.[1] These projects were anticipated to create thousands of jobs concentrated in construction and manufacturing. While labor-community coalition building was prioritized through policy implemented during the Biden administration, this model stood in opposition to the economic development agenda historically espoused by the city's business establishment. Atlanta, explained a former president of the state AFL-CIO, is a "corporate city" where "business is king."[2] Although unions participate in the public-private negotiation of regime politics in the city,[3] their power is constrained by Georgia's right-to-work laws that make it more difficult to maintain union membership and are a signature of the state's "pro-business" values. Today, fewer than 5 percent of workers in Georgia are union members.[4]

In part due to these legal impediments, unions have remained staunchly supportive of already-organized fossil infrastructure. A panel in the afternoon discussed the importance of jobs in coal-fired plants owned and operated by Georgia Power for union members. A union leader described that

> when we talk about closing down a fossil fuel plant, we lose 4-million man hours . . . That's a lot of people, that's gonna lose a job. . . . We can see it, we know it's coming. We understand that. We're behind it. We have to change the *mentality* of that working man or woman, that clean energy, new energy is a better alternative for all of us. But for them, to be honest with you, what they see is they're losing their job. That's a challenge. I'm not gonna say it isn't. . . . how [do] we address that with clean energy? We provide more jobs in clean energy than they had in the coal

> burner, right? We're not doing that yet. . . . We're not producing enough jobs in clean energy to replace them. That's why you'll watch us sometimes fight to keep places open because we're not ready yet in the state of Georgia, like we think we are.[5]

Yet, the US Energy and Employment Report produced annually by the DOE tells a slightly different story about how energy employment is changing. In 2023, solar alone employed nearly eleven times more people than coal-fired power plants.[6] The issue is not the number of jobs, but that unions have struggled to gain a foothold in the low-wage solar industry and have often strategically focused on maintaining high-wage coal and nuclear jobs preferred by their members rather than attempt to organize solar workers.

A growing literature on environmental labor studies has investigated organizing strategies and constraints unions face in industries in transition to build worker power.[7] Unions are democratic institutions that workers organize within workplaces to negotiate conditions on the job, wages and benefits. Benefits include health insurance, worker's compensation, disability, and a retirement package that provide care for workers and their families during and after their career.[8] The Building Trades unions, who represent the skilled craft workers in the electricity sector, also coordinate with employers to manage and fund apprenticeship programs and thus contribute to the social, political, and educational practices that define and maintain their member's skill premium. Blake, an IBEW leader, described his motivation as an organizer is to make sure all jobs are "livable wage jobs, where people can own a house and a vehicle and get back and forth to work and have a budget where they can still enjoy the eight hours of recreation that's what we all work for."[9] In bringing together workers to negotiate to improve the conditions of working and nonworking life, unions also build political power. Timothy Mitchell attributes the rise of working-class democratic power in the early twentieth century to organizing around the extraction and movement of coal in mines, dockyards, and power plants where hundreds of workers reported to work daily and could control concentrated chokepoints along supply chains. Strategic, collective action through strikes could force employers to negotiate.[10] Workers demanded high wages and benefits to offset the risks they faced in dangerous and dirty worksites. In contrast, the solar industry emerged in the early days of neoliberalism as an "unconventional" energy source that strived to usurp market share from incumbent energy sources by driving down costs, including through lower wages. Solar jobs are dispersed, and most are in construction, which are temporary in nature and often pieced out to multiple subcontractors. Outside of California where most utility-scale solar is union-built, the industry is majority nonunion.[11] The IBEW in Georgia estimated that 5–10 percent of solar work in Georgia was completed by union members in contrast to 15–20 percent representation in the traditional utility sector, which remains the most heavily unionized private-sector industry nationwide.[12] The question for unions then is perhaps not how to create more jobs, but how to organize the solar and broader clean energy industry to ensure that it creates *good jobs*. This is necessary to achieve what labor unions have described as a *just transition*, which would provide social protection for dislocated workers, such as building pathways to retirement or new employment for fossil fuel workers, and ensure the rights and protections of union jobs in the clean energy economy.[13]

BIL and IRA attempted to escape the zero-sum politics that pit jobs against environmental action by incentivizing a range of zero-carbon technologies, residential energy improvements, and high labor standards across the clean energy supply chain. For the first time, tax credits for clean energy that were extended under the IRA included a bonus for taxpayers who pay prevailing wages and hire apprentices to invest in the next generation of the clean energy workforce. These labor standards encouraged taxpayers to contract with unions who operate most apprenticeship programs and, explained IBEW 613 business manager Kenny Mullins, "are creating a new generation of good-paying union jobs that are good for our communities and our planet."[14] Manufacturing workers at the Blue Bird electric bus factory in Macon, Georgia, also leaned on federal policy to call for a union election. This successful vote to join the United Steelworkers defied the anti-worker development agenda rooted in anti-Black racism in the South as workers turn "to unions to balance the scales," and unions join with community partners "to fight for justice on all fronts—economic, environmental and racial."[15] If BIL and IRA buttressed the fragile groundwork for building a climate-labor-racial justice alliance by encouraging good jobs in publicly subsidized clean energy industries, actually moving away from fossil fuels as part of a real energy transition remains more divisive.[16]

At the end of the event, a long-time energy advocate stood up to thank the union leaders for starting a necessary and "candid conversation." While unions and environmental organizations have partnered on campaigns in

Atlanta, they have historically diverged when it comes to Georgia Power. Environmental and consumer advocates have militated to close coal plants, build solar, and oppose the $30 billion price tag for Plant Vogtle that they argued will exacerbate energy burden in the state. Unions have advocated at the Public Service Commission to keep coal plants in operation and finish construction at Vogtle, both sites where their members work. These patterns repeat across the US where Building Trades unions have been criticized as willing partners in the environmentally destructive growth mandate of electricity capital.[17]

In this chapter, I study how energy workers organize electricity employment, including through the partnership between Georgia Power and Atlanta's Building Trades locals who represent power plant employees, line workers, and construction workers who build and maintain the company's facilities. Electric life depends on the labor of Georgia Power employees, and their unions who have cosigned the utility's growth mandate to secure jobs for their members. On and off the job, I observe the unevenness of this relationship and how the utility captures value in myriad ways that control workers time such as requiring workers to be on-call while off the clock; employing racial wage scales that misclassified and underpaid workers of color relative to white workers; asking Black workers to represent the utility in their communities and mobilize their identities to build trust in the utility; and weaponizing the threat of outsourcing to nonunion contractors to exploit wage differentials between "skilled" and "unskilled" labor. Over time, workers have organized through unions and legal action to protest the exploitation and devaluation of their time. In taking stock of these strategies, as well as virulent anti-union state policies and the growth of nonunion clean energy industries, I argue that an analysis of the labor of electric life must consider both unions' relationship to the utility and the solar industry. While sympathetic to the constrained choices that unions undertake to protect their members, jobs, I advocate for an expansive approach to just transition committed to aggressive organizing of nonunion sectors that could combat the segmentation of the labor market, expand support for the energy transition within union membership, and broaden mobilization for clean and affordable energy as part of a working-class environmentalism committed to a social wage.[18] In focusing on labor organizing in the social reproduction of the workforce, I bring a feminist lens to expand limited studies of labor within urban political ecology.[19]

Beginning with a history of union organizing in the construction sector in Atlanta, I trace how work in the electric utility was constituted—or articulated, as Stuart Hall defines it—alongside and through an accreted racial, gender, and class division that emerged between "skilled" workers and "unskilled" or "deskilled" laborers.[20] When Black workers sued Georgia Power and the union over discriminatory policies during the civil rights movement, the utility used diversifying its workforce as a reason to move away from union contractors. Yet, allegations of discrimination continued across union and nonunion blue- and white-collar jobsites at Georgia Power. The bifurcated labor geographies that emerged, differentiating higher-wage, unionized workers and lower-wage, nonunion workers map onto energy technologies as seen in contrasting the utility's recent investments in nuclear and solar. In tracing this history, I argue for the central importance of building worker power as part of an energy democracy movement representative of and responsive to the needs of working people and committed to creating good jobs across the economy as part of a just transition.

The Company Union

The history of unionism at Georgia Power mirrors a national trajectory of organized labor in the early twentieth century. Moving from an early opposition to unionism, the utility came to partner with its unions who aided in reproducing a segregated workforce and political consent for the company's grow-and-build model.

In Georgia, labor organizing predates the utility and began in the antebellum period with the collective efforts and flights to freedom of enslaved Indigenous and African people, indentured servants, artisans, and sailors.[21] The union movement emerged in the nineteenth century from workers' lived experience to challenge enslavers' and employers' control over time. As a tool of occupational organizing, unions were also used to contest and police the color line. Before the Civil War, white workers organized to restrict competition from free and enslaved Black craftspeople.[22]

After the Civil War, unionism grew in Atlanta, which was the biggest railroad hub in the South and a growing industrial center. Notably, Black women led one of the earliest mass labor actions in the country in the Atlanta Washer Women's strike in 1881.[23] Building Trades unions also organized contractors and worksites to create a monopoly on skilled labor and maintain their bargaining position by limiting the number of workers who had access to apprenticeship, licensing, and union membership.[24] Black workers in Atlanta constituted a large share of the carpenters, brick and stone masons, painters and glaziers, and plumbers in unions, and a significant majority of the plasters and cement finishers, however, their share of electrical work fell from 6.5 percent in 1890 to 0.1 percent by 1930.[25] Across the trades, the share of Black workers represented in construction unions began to fall with the rise of Jim Crow as white union members struck or boycotted employers who tried to integrate worksites and, in some incidents, violently barred Black workers from railroad, maintenance, and construction jobs.[26] A survey conducted by Walter Westmoreland at Atlanta University found that "the persistent and strenuous opposition of the white unions [compelled by] economic and race motives cause them to, by secret or open agreement, to oppose any effort of Negro mechanics to become an integral part of the local labor movement."[27] Black labor organizers and academics cautioned that racial antagonism would undermine the labor movement as employers used race to classify and divide workers and denigrate unionism.[28]

Despite growing unionism in Atlanta, the utility maintained an anti-union shop. Atlanta Mayor James Key reportedly joked that if a utility company employee "even walked through the Union Depot [railroad station] he lost his job."[29] The utility "only hired white men raised in the country as motormen and conductors, in part to get employees with no union background."[30] A former streetcar driver, A. J. Sharpe, who had migrated to Atlanta from rural Georgia, remembered a massive hiring day in which hundreds of men gathered to jockey for a job as a distinct break from farming and dairying. This competition between prospective workers benefitted the utility. Another employee recalled that for drivers who were just beginning without a regular shift, "Three times a day you would report [to the barn to see if someone had called in sick;] but didn't get any time for reporting. If you did work, it would be around eleven or twelve hours a day." He joked "that children asked, 'Mama, who was that man in here last night?' . . . You couldn't hardly spend no time with your family hardly at all."[31] Workers began to protest the nonstandard shifts, which enrolled unwaged time that workers spent waiting to be scheduled to maintain a pool of surplus workers and depress wages across the board.[32]

Economic pressures during World War I catalyzed organizing. With wartime inflation mounting, electrical workers walked out in 1916 to protest the declining value of their paycheck and called in the IBEW to lead the strike. They were followed soon after by the streetcar workers who joined the Amalgamated Association of Street and Electric Railway Employees. The dispute between the company and employees was not resolved until the Company conceded to recognize the Amalgamated Association Local 732 under the conditions laid out by the National War Labor Board. Although it maintained an open shop, 90 percent of operators joined the union.[33] The Amalgamated Association struck again in 1920, demanding standardized schedules and pay increases. In turn, the Company requested better conduct from employees. The terms were formalized in a memorandum of understanding that "united the company and its workforce in a partnership dedicated to public service" until the utility sold its transit arm.[34] In 1919, the utility recognized the IBEW Local 84 as the representative of linemen and power plant operators. The utility also utilized union contractors, who employed boilermakers, pipefitters, sheet metal workers, operating engineers, and others at its power plants.[35] Georgia Power would go onto employ more unionized construction workers than any other company in Georgia over the next two decades. While the relationship between the utility and union was not always amicable—for instance, the IBEW struck in 1930—both parties mutually characterized their relationship as one of "friendly co-operation."[36] As the builders, linemen, and maintenance and operation workers interested in growing their share of work, the union's mission aligned with the utility's growth ambitions.

The unions became advocates on behalf of the utility's grow-and-build model. In one of the earliest instances, the Atlanta Federation of Trades lobbied for the city ordinance that prohibited, among other private transportation services, the buses operated by Black neighborhood associations to avoid Jim Crow seating on the streetcars.[37] Arkwright fondly spoke in support of the IBEW. At a speech before the Southern Labor banquet that he hosted, he expressed that "the union has proven beneficial to the Georgia Power Company . . . It has resulted in the establishment of confidence and good-will between employer and employe [sic], and I hope to see the day when we will have 100 per cent [sic] union membership among our employes [sic] in Atlanta."[38]

Following the 1935 adoption of the National Labor Relations Act (NLRA), which affirmed workers' rights to organize, bargain collectively, and strike, the number of Georgia Power employees who were IBEW members increased as the union expanded outside of Atlanta and to include entry-level and service workers who were predominately Black. Yet, contracts reproduced a racial hierarchy in titles and wages. Laborers were added to the electrician's bargaining unit in 1943; janitors, porters, maids, and meter readers were organized as separate sections in 1945 and 1947. The contracts for the Amalgamated Association at first strictly limited Black workers to "colored" positions, a category changed to "laborer" in 1943. Laborers were paid less than half the hourly wage of operators and mechanics and received fewer wage increases over contract negotiations. The contracts also reinforced the color line in occupational duties: Black men could not be substituted into "work regularly performed by white men" with the exception of emergencies or "where assistance is required on heavy work."[39] In the IBEW where "the backbone of the organization is based on skill," skill also became a designation to diminish the talents and capacities of Black workers who were either not provided or excluded from formal training.[40]

The NLRA worked alongside other New Deal programs such that "collective bargaining became a centerpiece of the nation's social welfare regime."[41] Benefits for nonunionized workers also improved as employers in major industries kept pace with union wage gains. However, the negotiation of living standards through collective bargaining supplanted a comprehensive federal system to secure a social wage available to all that would not depend on wages.[42] For most people, economic security was (and remains) incumbent on current or past labor within a hierarchical labor regime. Organizing wages and benefits through employment gave rise to "a stratified system" and entrenched a racialized and gendered division of labor both within a company like Georgia Power and across the economy.[43] As Du Bois presaged in *Black Reconstruction*, published the same year that the NLRA was adopted, a bifurcation of labor between skilled trade unions and "common" workers emerged and the unstable alliance between capital and labor formed at the expense of unorganized workers began to erode within a decade.[44]

The most damaging blow to the labor movement was the 1947 Taft–Hartley Act, which brought into force right-to-work laws in the South, pilloried by unions as "the right-to-work for less." These laws banned union security agreements or "closed shops." This meant that unions had to represent any employee working within a bargaining unit, but workers could not be compelled to pay dues to fund legal assistance, organizing, and other staff, creating a "free rider" problem for the union. The Southern Industrial Council, among other industry groups in the region, had propagandized to advance right-to work, arguing that union security agreements "put the individual worker under a system of enforced servitude."[45] Despite Arkwright's prior enthusiasm for the IBEW, Georgia Power's official history of labor relations describes that Taft–Hartley "validated Georgia's Right-to-Work Law" and a regional way of doing business.[46] Right-to-work laws diminished union bargaining power and efforts to increase union density in the region. The anti-union atmosphere became a defining feature of the pro-business environment in the South.[47] Labor leader A. Philip Randolph described Taft–Hartley as a geographical problem of racialized uneven development. Labor-friendly states lose jobs with industries moving to "low-wage non-union states," and "the black and white working poor in the 'Right-to-Work' states of the South, noted for sweat-shop starvation wages, are not free to organize bona-fide unions to fight for living wages."[48] Despite impressive gains between 1930 and 1950, when union membership increased from roughly 7 percent to more than 30 percent nationwide, union membership in Georgia never exceeded 11 percent of the workforce.[49]

Unionism is one way in which workers join together to bargain collectively to improve their lives on and off the job.[50] In the face of racism and sexism at the workplace, unions have been especially important for women and people of color to fight discrimination.[51] Beginning in the 1960s, BIPOC workers and women used civil rights law to challenge discrimination on the job and within unions and took collective action, including striking at worksites across Atlanta.[52] At Georgia Power, civil rights actions by Black union members disrupted the racially coded division between skilled workers and laborers.

Questioning the Atlanta Way

When the utility recognized workers' bargaining rights in the early twentieth century, the workplace remained segregated along occupational lines. The use of racism and sexism to organize labor became another means to profit in systematically underpaying workers as the utility grew. A deeply unequal workforce in which a worker's race determined their opportunity for career advancement powered electric life. During the civil rights movement, Black Georgia Power employees used Title VII of the 1964 Civil Rights Act, which prohibited employment discrimination, to challenge occupational segregation. Reviewing the testimony of Black workers who used lawsuits, collective action through groups such as the NAACP, and the press to protest discrimination in opposition to the conciliatory politics of the Atlanta Way that accepted slow progressive movement toward racial equity reveals the ways in which workers identities, as much as their labor, sustained the legitimacy of the utility as it sought to cultivate a culture representative of all Georgians.

The first action against Georgia Power came in 1966 when five Black workers filed a complaint with the federal Equal Employment Opportunity Commission (EEOC). A year later, the EEOC reported "'reasonable cause'" to believe the company and union were in violation of the Civil Rights Act. After the EEOC was unable to broker voluntary compliance, Black workers sued the utility in 1968 because it "classified [them] out of the line of progression in all departments which restricts them to laborer posts and limits their promotional opportunities." The lawsuit also charged that the utility maintained "segregated comfort facilities."[53] Of the 450 Black men that Georgia Power employed, 350 were laborers and 97 percent of workers classified as laborers were Black. IBEW Local 84 was named as a codefendant given its role in maintaining a promotional system that limited opportunities for Black workers. The union contract required laborers to pass a test or meet a minimum education requirement to be promoted, while promotion from other job classifications was based only on seniority and competence.[54] The test was not about ensuring that workers had necessary skills to succeed in higher-level positions, but rather maintaining the bifurcation of the workforce that made it almost impossible for Black workers to advance. One plaintiff, Mr. Charles King, was not promoted after being told that he passed the aptitude test twice.[55] Another plaintiff, Mr. Willie Moreman, described that white workers hired as laborers were quickly moved into "more desirable positions."[56]

The racialized hierarchy of occupational titles obscured misclassification of Black workers who were required to perform a variety of tasks outside of their contractual job duties but underpaid for their work. In affidavits submitted as part of the case, workers testified to the dangerous and onerous working conditions outdoors, at Plant McDonough, on switchyards, and on the coal belt, termed "the hole," that they were required to take on despite being classified and paid an hourly rate of $1.82 as laborers.[57] Mr. John Swanson wrote, "I feel that I am being discriminated against as a Negro because this job was not reclassified from Laborer to Switchmen" even as he assumed the responsibilities, duties, and risks of the "very dangerous" job where "one mistake could mean loss of life, loss of one or both arms or legs." The union told him that the utility would not reclassify his job but would offer a $20 increase in pay.[58]

In 1971, a court ruled against Georgia Power, requiring the utility to offer Black job applicants who had been refused positions or promotions "the first available jobs for which they are qualified." The same week that the case was closed against Georgia Power in federal court, women employees sued the utility for gender discrimination.[59] Three years later, a US District Court issued what was at that point the largest discrimination settlement recorded in the US and required Georgia Power to pay $1.75 million in backpay to 360 workers, as well as pension benefits denied to those who had retired or were soon to retire. The use of non-job-related aptitude tests and educational requirements was prohibited and the utility mandated to meet hiring quotas to increase the share of Black workers from 12 percent to 22 percent in the electrical union, 9.6 percent to 22 percent in the clerical workforce, and less than 1 percent to 6 percent in managerial positions.[60] At the time, only 0.5 percent of IBEW members in Atlanta were not white.[61] Despite these charges, the history of utility labor relations authored by a former Georgia Power vice president asserts that "the Company did not and does not, to this date, acknowledge guilt as charged by the Justice Department."[62] In 1982, this case reached the US Supreme Court, which ultimately voided the lower-court decision that Black employees faced discrimination "by a union-approved seniority system."[63]

Court action against Georgia Power was buttressed by the Department of Labor, which imposed the Atlanta Plan to increase the share of workers of color who were members of ten out of seventeen construction trades.[64] At the time, Black, Latino, Asian, and Indigenous workers represented only 7 percent of Atlanta's Building Trades members despite comprising more than 23 percent of the metro's population.[65] Working with organizations such as the Urban League, the Building Trades committed to train "semi-skilled and unskilled individuals to become journeymen craftsmen."[66] The city also established an Equal Employment Opportunity Office in 1970, which languished until the election of Mayor Maynard Jackson. Among his most impactful programs, Mayor Jackson set requirements to contract or subcontract 25 percent of the city's airport expansion to minority-owned businesses.[67] The share of certified, minority-owned firms working in construction, engineering, and architecture increased, as did the share of Black workers in majority-Black construction workplaces. Black contractors successfully used their "political base as an interface to influence the structuring of the labor market in government employment and construction," winning bids and securing good jobs, including in unionized firms, on city-connected projects that changed the structure of contracting in the city.[68]

Despite gains through affirmative action, complaints of racial discrimination within white- and blue-collar work at Georgia Power continued. In the early 2000s, four workers brought a lawsuit alleging they were terminated forty-eight hours after reporting harassment and offensive working conditions that had caused one hundred Black and Latino workers to quit construction jobs at a power plant.[69] In 2000, another suit was brought by seven company employees to win class action status to be able to represent all 2,400 Black employees affected by "systematic" racial discrimination.[70] They charged "'reckless indifference' to a racially hostile workplace" for African Americans,[71] as well as racial and gender inequities in pay, reviews, and promotion, including "abandoning" seniority measures for promoting foremen and using qualification exams with "'a disparate impact on African American employees,'"[72] in ways that continued to devalue the time of Black workers. The *Atlanta Constitution* positioned the lawsuits alongside charges filed against other Atlanta-based companies including Lockheed Martin, Home Depot, Waffle House, Coca-Cola, and Bell South as part of a push for "advancement and promotion" in the next phase of "civil rights struggle" in employment.[73] Complainants insisted that despite the downward effects suits might have on "profits and morale at large companies," Black workers deserved more than entry to the workplace and should be "remaining and thriving" in corporate Atlanta.[74] Georgia Power Vice President Jim Davis, who had been hired following the 1968 case to lead the human resources division to integrate the Company, expressed disappointment that employees sought to litigate their complaints rather than use the channels provided by the utility,[75] and that despite some efforts, Black workers continued to experience what one litigant named an unchanged culture of the "good ol' boy network."[76] Rejecting the conciliatory politics of the Atlanta Way managed under the biracial urban regime, working-class Black workers laid bare the harassment, intimidation, and underpayment they experienced on the job and criticized the devaluation of their work that undergirded company profit.

In an internal review of their hiring and promotion records revealed through the lawsuit, Southern Company found that Black workers were more often fired and less often promoted. Ms. Patricia Harris, a market research analyst who quit her position in frustration that she was less valued than her white coworkers with less experience, told the *Atlanta Constitution* that the data on promotion trends within the utility confirmed her experience was systemic. She described: "I just felt stripped. I felt worse. I had worked very hard for the company, and I'd represented the company wherever I went. I'd always come back to my hometown during career days, and talk it up."[77] Following the Company's mantra of being a "a citizen wherever we serve," Harris had represented the utility throughout Georgia only to realize that she and other Black workers who faced experiences of racism and sexism at the workplace ranging from harassment to violence and dismissal were also underpaid. Her racial identity, experience, and affective labor became part of what Georgia Power was selling as she was asked to represent the utility in Black communities and build its public image. Ultimately, the suit was dismissed in court. Of the seven workers who challenged the utility, one worker was fired and others who had already quit were not able to see the changes implemented within Georgia Power that their concerns and action helped generate.[78]

Public efforts to hold the utility accountable to representing and respecting workers from across its service territory contributed to its rebranding as a multicultural institution that proudly promotes its accolades as a "top company for diversity and inclusion."[79] The annual Southern Company Corporate Responsibility Report publishes general information on racial and gender diversity. Recruitment and hiring of nonwhite people has been a particular priority for the Company.[80] In 2022, 26 percent of its workforce were women, including 25 percent of management, and 30 percent are people of color, including 23 percent of management.[81] Notably, approximately 50 percent of Georgia's population is nonwhite. White men continue to be overrepresented relative to Georgia's population among craft workers at Georgia Power and in the broader energy industry.[82] In 2022, Chris Womack became the first Black person appointed chairman, president, and CEO of Georgia Power. He was subsequently promoted to president and CEO of Southern Company the following year. At the time of his appointment, Womack was only one of eight Black CEOs of Fortune 500 companies.[83]

Without diminishing these efforts to improve diversity, inclusion, and equity at work, it is also necessary to ask how the promotion of multiculturalism within the utility falls short in addressing the routine effects of racialized uneven development reproduced over time and through uneven access to affordable electricity.[84] While the utility works to reverse course on decades of discriminatory hiring practices that contributed to the devaluation of Black, Latino, and women worker's time with the political cooperation of the union, repairing the ways in which the promotion of electric life has reproduced hierarchies of power and difference is a much larger task.[85] The energy transition threatens to disrupt the co-eval organization of electricity, labor, and the built environment but this shift will not necessarily repair or improve outcomes for working people, as a comparison of between nuclear and solar industry organization reveals.

Union Power

When the first lawsuit concerning racial discrimination was filed against Georgia Power, the utility was also negotiating a shift in its contracting process that would alter the half-century partnership with organized labor. Unions had supported and benefitted from the utility's grow-and-build agenda. Over time, a division of labor between organized and nonunionized workers distinguished by technology emerged and this gulf would inhibit solidarity organizing between organized labor and environmentalists.

In 1970, the utility spent $368 million for construction, exceeding all other industrial construction in the state that year by more than 50 percent, and had an additional $2 billion in construction planned. A report in the magazine *Southern Exposure* described: "Georgia Power virtually controls the construction industry, and with it the wage scale of thousands of workers,"[86] and "The big jobs have always gone to contractors using union labor."[87] Facing a plateau in technological improvement, new environmental regulation and the energy crisis that increased fuel and financing costs,[88] Georgia Power proposed hiring a nonunion contractor to build a new $750 million coal-fired power plant southwest of Atlanta. In doing so, the utility mirrored a position advanced by the Nixon administration and other executives through the Construction Users' Anti-Inflation Roundtable to blame rising inflation on the high wages of construction unions, beckoning a "policy shift away from union labor [that] is part of a larger attempt to thwart the power of construction workers."[89] The utility further targeted the union as the source of racial discrimination, citing that a nonunion contractor would increase the share of Black workers employed from 5 percent to 15 percent.[90]

The issue was existential for unions who worried their members' future when a massive construction budget was on the table. An officer with Atlanta IBEW 613, a construction local that has a long history of performing contracted maintenance and construction work for Georgia Power, described he was trying "to convince the power company that union labor can do a better job."[91] Adding in correspondence to the *IBEW Journal* that:

> There are times when this correspondent becomes as confused as a blind man in a dark room chasing a black cat that may not be there. . . . IBEW Local 84, Atlanta, is the bargaining agent for most of the [Georgia Power] employees. IBEW Local 84, Atlanta represents the employees of contract line construction doing business with the company, and several inside IBEW locals are employed on inside construction, the large majority of which is generating plants, on power company property. . . . But what happens now? The Georgia Power Company has, so far this year, contracted—with non-union construction firms—200 miles of high-voltage transmission lines, some warehousing and office facilities, and a huge multi-million-dollar generating station. How long can such a dual operation be endured?[92]

In response to what felt like a violation of the labor-management compact, Building Trades members walked out on wildcat strikes across the state, picketing construction sites, and dynamiting and shooting electrical substations.[93]

The strike galvanized support from a broader coalition in Atlanta. An exposé published in *The Great Speckled Bird* offered a nuanced analysis of the Building Trades position. As a former union lawyer explained: "They're one of the most exclusive groups going. They've made it, and they're not about to let others, like Black [worker]s in. There are better things to support than their strike." Yet, the publication also militated for workers' right to organize, noting that "if Georgia Power really cared about getting Black [worker]s good jobs, they could launch a massive program to hire Black [worker]s in all levels of the company" and design "an 'affirmative action' plan,"[94] to get the company and union to integrate more quickly. The Georgia Power Project ultimately supported the strike.[95] Simultaneously, the Atlanta Labor Council, representing Building Trades, industrial, and public sector unions, opposed the rate increase that would finance construction at issue and mobilized members to speak out at the Public Service Commission (see chapter 2).[96] This brief alliance of cross-cutting opposition to Georgia Power between consumer advocacy and union movements, was a moment of possibility.[97] However, the movement was not enough to change the business practices of the utility, or the union that depended on the utility for work. Ultimately,

Georgia Power secured funding to continue some of the construction it had planned, although its nonunion contracts also endured.

The utility and unions reconciled on a wave of nuclear expansion, signing a collective bargaining agreement to cover an initial $2.6 billion investment to construct Plant Vogtle in Burke County that would create ten years of work.[98] At the 1975 regional IBEW meeting, Georgia Power President Scherer "commended the IBEW for the position it had taken regarding the recent request of the Georgia Power Company for a rate increase."[99] In so doing, Scherer recognized workers for their labor beyond their paid work on the job, noting the importance of the organizing capacity of the union to mobilize workers to speak in support of expansion at the Commission.

Despite the partnership between the unions and the utility on Vogtle Units 1 and 2, the expansion of Units 3 and 4 started as a nonunion job. The Augusta Trades and Labor Council formed the "Jobs for Georgians" campaign "to demonstrate that using skilled union trades labor was the best option for building the reactors."[100] Unions heavily lobbied the Obama administration, which extended credit to build the facility through the DOE Loan Program Office, to ensure the company would hire union contractors, raising concerns about poor quality workmanship associated with reactor pieces that were prefabricated offsite in a nonunion shop.[101] Southern Company and its contractors ultimately signed project labor agreements covering the site and future nuclear construction.

Job creation became a central rationale for the project as the timeline and budget ballooned, alternatives like solar became less expensive, and energy efficiency improvements meant that overall demand for power flatlined between 2008 and 2022. The utility justified continuing construction based on the "5000 jobs [at the site]. Governor Deal and the Commissioners are . . . [saying] 'not on our watch are we going to lose all these jobs,'" shared Beth, a consumer advocate.[102] Where the project was once billed as a way to meet power demand, "Now, it's all about jobs" explained Hugh in 2019. The utility would brag: "This is the largest construction project in Georgia. . . . an economic engine of Southeast Georgia."[103]

As costs mounted, Building Trades leant their support at the Commission for continued authorization of the project. The state AFL-CIO endorsed supportive political candidates for Commission seats, including Republicans, in contrast to the usual party line support organized labor threw behind Democrats. An IBEW leader attributed the perhaps unexpected endorsements to the fact that "the dignity that comes from having a job is more important than any politics."[104]

Yet, Vogtle would remain a contentious dividing line between unions and environmental and environmental justice advocates who raised concerns about the added cost of nuclear that exacerbates energy burden in low-income communities across the state and the risks to environmental and occupational health along the nuclear fuel supply line.[105] Located in a rural, agricultural community on the South Carolina state line, the plant "happened to be in one few rural Black neighborhoods in the county so it started off as an economic justice problem, and remains that," explained Liam.[106]

The Building Trades have also supported the continued operation of coal- and gas-fired power plants. Under a national maintenance agreement signed with the Southern Company in 2020 that builds on the nuclear project labor agreement, construction unions and signatory contractors are contracted for maintenance work at existing plants and new construction work on fossil facilities. IBEW Local 84 has jurisdiction over maintenance and operation work, transmission, and distribution and represents electricians at two coal plants, two nuclear sites, three of the six gas facilities, and several hydro facilities.[107] Beth shared that her organization at times reached out to the unions to try to find common cause on issues, but "overwhelmingly the unions are going to side with Georgia Power." She added, "Georgia Power does in fact employ a lot of Georgians and I don't mean to suggest that, that's not important, that there are good jobs," but it felt disingenuous to her not to acknowledge that "there's a lot more jobs in clean energy."[108] There might be more jobs in clean energy, but as noted at the top of the chapter, these jobs are less likely to be unionized.

At the same time, Building Trades unions have fostered partnerships with Atlanta community organizations. Apart from the corporate initiatives to advance diversity undertaken by the utility, the Atlanta Building Trades and state AFL-CIO helped to found Georgia STAND-UP, a "think and act tank" for working families in Atlanta in 2004, which sponsored, among other initiatives, the apprenticeship readiness Trade-Up program.[109] Georgia STAND-UP director Deborah Scott wrote that the program aimed to build trust between "two historically suspicious interests—organized labor and the African American community—marginalized within a region inimical to both [to identify . . .] common ground and mutual interests around which to forge a functional partnership." Labor benefitted in being able to diversify its aging and "overwhelmingly white workforce," and Black community organizations saw opportunities to support un(der)employed workers in improving their "economic fortunes."[110] In 2011, Georgia STAND-UP also helped convene labor, unions, and environmental organizations as part of the Emerald Cities Collaborative, a national organization chartered by the Building Trades to build a "social compact between labor and community," and to build interest in clean energy jobs.[111]

Since 2011, when construction at Vogtle started, the fuel mix at Georgia Power has shifted from coal to gas and solar, which requires a smaller permanent workforce. Workers at closing coal facilities were offered early retirement packages or the opportunity to transfer to other plants.[112] While the utility has not fired anyone, there is no affirmative commitment to expand the union contracts into new energy resources.[113] In an interview, one retired union member admitted, it had been hard to get local Building Trades to pay attention to clean energy. Notably, their leadership had failed to follow through on the Emerald Cities effort to bring together downtown real estate owners, the city, and unions to sign project labor agreements on large commercial efficiency retrofits during the financial crisis, when Atlanta began to implement the Better Buildings Challenge.[114] Another labor official commented that when the maintenance agreement with Southern Company was negotiated, solar jobs, which are often lower paying, were not the priority.[115] Just as unions have failed to be more proactive on organizing clean energy jobs, the utility and Commission have also failed to take up the question of a just transition, at least in any public forum, to build a pathway for incumbent workers from fossil fuels to clean energy.

While union members have built broad coalitions to promote quality jobs for workers marginalized within the Southern economic development agenda, building trust between labor and environmental organizations has proven more challenging.[116] One energy advocate disparaged that the narrative of green jobs and retraining coal workers to put up solar panels revealed "a great naivete on the part of the green community, that we just think that it's that easy."[117] Another said more bluntly that workers were not considered by members of his organization who "did not really have working-class sensibilities."[118] While energy burden would become a central issue in Atlanta uniting environmental and energy justice groups at the Commission, most remained quiet on the future and quality of energy jobs with solar prioritized as the cheapest way to implement zero-carbon energy.

Installer Brotherhood

As the Vogtle expansion was under construction, solar power boomed in Georgia. Opposition from the utility to implementing renewable energy contributed to the emergence of bespoke solar contractors who started by selling their niche product to environmentally conscious consumers. As technology improved, costs declined, and interest increased, solar enthusiasts focused on making the case for solar to the utility based not on environmental or climate grounds, but on solar's cost savings. Solar contractors found themselves in a race to the bottom that accelerated the decades-long process of breaking the power of the union and reducing the security of the union contract for incumbent and future employees. Across the industry there remains an idea that solar installation work is relatively rudimentary, not requiring the skill of other kinds of construction or utility sector employment. Yet, this stigma further erodes the bargaining position of workers who are often hired through temporary or precarious contracts and disregards the expertise workers bring to the job. Nationwide, workers have organized to improve labor conditions in solar; however, in the South, unions and their contractors, who command a higher wage, have struggled to enter the field.[119] The utility is able to reap the benefit from lower-cost solar power made possible by making labor cheaper.

Over the last twenty years, the solar industry in Georgia has grown from mostly mom-and-pop residential solar installation firms to 250 companies engaged in manufacturing, installation, production, and design, who employ more than five thousand people. Since 2013, when the Public Service Commission mandated Georgia Power include solar in their fuel mix, more than 6.1 GW of solar generation capacity have been built with an additional 3.2 GW planned through 2029. Georgia has the seventh most installed solar capacity of any state in the US.[120] A small share comes from distributed generation solar on homes and businesses. The majority of this is at the utility scale and is sold directly to Georgia Power, other utilities, or companies such as data centers. Edward, a retired solar developer, proudly talked me through the evolution of his company from 2007, when "to build a solar company, we had to build a market. In order to build a market, we had to create an ecosystem."[121] As the ecosystem evolved, the solar industry focused on trying to prove its benefit to Public Service Commissioners and overcome opposition from Georgia Power by showing it was simply the cheapest option available that could outcompete coal, gas, and nuclear on cost.

Unlike in other states with supportive climate policy that incentivized solar, Georgia does not have a renewable portfolio standard in place. In interviews with solar developers working at a range of residential to utility scale, all felt constrained by the legislative and regulatory unpredictability that also fostered competition for work between companies. In 2019, Shawn, a contractor, pointed out that changing the "regulations could definitely [help] . . . just more incentives to create more customers, you know, we just want to work. We don't want to have to chase work all the time."[122]

Competition benefited the utility. Most of its solar capacity is procured through power purchase agreements with developers who hire contractors to build installations and then sell their energy back to the utility. Joe explained, "Some companies that are just like in this race to the bottom in Georgia. I mean, it's highly competitive. . . . If it's a significant project, there could be seven or eight different solar companies looking at this, and most of them are willing to have a very small margin . . . or even lose money on a project just to get the foot in the door with that customer to get the marketing rights for that project. And that's just. . . . that's been going on for years. And it's caused a lot of companies to go out of business."[123] Evan, a solar contractor, added the Public Service Commission is laser focused on cost and reliability. Taking other concerns into account could leave the Commissioners open to lawsuits if their decisions turned out to be wasteful.[124] Facing this narrow band of decision-making, Roy, a representative for a solar trade organization argued that "the way to win at the Public Service Commission was just to show that we're the cheapest option out there."[125]

Shawn, who started as an installer and worked his way up to own his own company, observed that the cutthroat nature of the industry often hurts employees who became

> the battering ram for them to get that [installation] done. . . . You think it's a green industry and everyone's . . . really good-hearted people that want to help customers and [also] realize their crew's out there in 100-degree weather driving all over the place—it's a really tough job—you think they would appreciate that, but the ones that I work with haven't.[126]

Other workers reported getting short-changed by predatory solar companies that went out of business before paying commission to their salespeople.[127] A Black solar installer described feeling unsafe after being dropped off to do an installation in rural north Georgia without a vehicle to leave.[128]

Wage theft and mistreatment on the job routinely reported by solar workers pattern labor violations in the broader Southern construction industry.[129]

Most solar jobs are in construction, and job quality varies widely. In Georgia, utility-scale work will "typically go to out of state contractors"[130] who often "come in and hire a lot of temp labor to get them installed."[131] Historically, contractors have an incentive to minimize training and cut down on labor costs. Being able to show up on time was the most important priority in hiring for many contractors I interviewed.[132] Training someone to do more complicated wiring is a six-week process.[133] Evan described that applicants from the few solar training programs in that state rise to the "top of the hiring list" and "if people can say—if they have roofing experience, or if they can say I went to tech school . . . for solar or to do electrical work," that is even better.[134] For the most part, explained Shawn, "We just we hire 'em and we train 'em. . . . You never know how good they are until they get out there a day or two. And you can tell [if] they're an all-star, so, who cares if they just got out of prison, or if they just got out of college?. . . . We don't have a problem with finding workers we have a problem with finding skilled guys."[135] From Shawn's perspective, among the skilled installers, "We all know each other. . . . The installer brotherhood is already pretty closed loop here."[136] Unlike the IBEW, the "installer brotherhood" is more diffuse, and employees rarely receive the same level of compensation or representation through a union to negotiate the value of their time in contingent work.

Finding "skilled guys" is a problem not only in the nascent solar sector but across the construction industry where job quality varies widely.[137] Hiring difficulty is also about the pay, benefits, and quality of job on offer. Low pay, dangerous working conditions, and long hours in all kinds of weather make it hard for solar firms to maintain a consistent workforce. Companies also admitted to exploiting a hierarchy of what they identified as skilled and unskilled workers. Edward described that on any jobsite you have "engineers that are certainly white collar, high paying jobs, but a solar installer that's just doing racking, or solar panel installation is going to make, you know, $15 to $20 an hour. Which is, depending on their skills, may be a good job." Racking and mounting panels is the most labor intensive because each panel is torqued by hand, but he continued, "It's construction. It's . . . at the end of the day, it's turning wrenches and drills and lifting things." Across these positions, direct employees usually receive benefits and "the temporary labor people will typically work for a temp labor organization. And the better ones of those would have benefits but you know, in construction, it's a mixed bag."[138] Edward's characterization was echoed by many contractors who described that solar installation is not skill intensive but labor intensive.

This dichotomy between "skilled," direct-hire positions and "unskilled," temp positions reinforces the Southern economic development agenda based on making labor cheap, diminishing the skills that workers bring and learn to survive on the worksite while reinforcing precarity and transience that limits upward mobility in the industry. Often, marginalized workers are the most exploited. The most comprehensive study of the solar workforce to date in Texas found large disparities in pay between white men, Latino and Black workers, and women doing the same job, with few workers having the benefits of a contract or union representation.[139] While all construction work is temporary in nature, unions provide portable health insurance, pensions, disability, and worker's compensation, and continued investment in apprenticeship training that contractors pay into as part of their hourly wages and benefits. This model ensures that despite the contingency of construction jobs, workers, retirees, and trainees have some security through collective risk-sharing in the absence of a social wage.

Despite these benefits for workers, unions struggle to organize solar in Georgia.[140] Edward stated flatly that at "the scale of solar projects, collective bargaining is not really something I've ever heard of. In certain states, but not in Georgia."[141] A developer dismissively derided an IBEW business manager in an interview for trying to negotiate for prevailing wage.[142] Before the IRA was adopted, only one solar developer I interviewed was signatory to the union and admitted it did not always benefit his company because not all union members he would hire from the union hall had solar experience. Another explained that she tried (unsuccessfully), but she felt as though she was not taken seriously as a woman, asking, "Are they ready to talk to me?" Her conversations stalled, as she remembered, because "the unions just didn't think that they had a dog in the fight. . . . Maybe they're just afraid of change, maybe, and they do have a lot of accounts with Georgia Power, so, they feel like that would be crossing the line or something? And I get it, union guys, they're loyal. . . . But Georgia Power [is] administering these huge utility scale projects for the IRP. . . . So, where's the union component in that?"[143] From her perspective, the utility was not representing the interests of unions; however, trying to hire union members became another obstacle to break into the cutthroat, very male, and very white solar industry.[144]

The IRA changed the calculus for solar contracting both by stabilizing the industry to reduce the cutthroat race to the bottom in prices and by incentivizing employers to invest in registered apprenticeship programs and improve pay by requiring prevailing wages.[145] Luke, a labor leader, described that the IRA provides "a little bit of encouragement" because "that is federal money coming to Georgia and so therefore, we can go after prevailing wage, and also enter into project labor agreements with the utility companies on these and that will help even for the solar installations that are part of this program. It'll help lead them towards union contractors and union jobs. But if it doesn't have federal funding, you're dealing with day laborers pretty much."[146] Despite these generous incentives, federal grants and tax credits do not guarantee work for the unions, and outcomes for union members have been mixed. While some have praised the tax credits, citing new interest from contractors, union membership in the private construction industry nationwide and in the Atlanta metro area fell to a record low in 2024.[147]

Solar developers have devised other strategies to comply with the IRA. Kevin, an apprenticeship training coordinator for a solar developer, explained that there is no way a company would build a project without going for the enhanced tax credit for improved labor standards. It's the "difference between 6% and 30."[148] His company works across the country. In pro-union states, they often subcontract to union contractors. Yet, he flatly said, it's not how they do work in the South. The company has registered their own apprenticeship program to get the enhanced tax credit instead. As the labor market tightened, they also pay well above the South's low prevailing wage levels to attract workers. This trend has been adopted by many of the largest solar companies.[149] Other contractors have joined nonunion, group apprenticeship programs that provide required supervision and on-the-job training while a consultant covers classes and registration paperwork.[150] Despite its partnership with organized labor at its own facilities, Georgia Power promotes non-union apprenticeship programs and education and workforce development programs as part of its economic development arm that recruits companies nationwide and internationally. Apprenticeship offers a credential, but the benefits of union membership, such as a voice on the job and portable benefits, are not guaranteed.

State politicians in Georgia have responded to federal incentives by ramping up anti-union rhetoric and legislation. In April 2024, as workers at electric vehicle plants in Tennessee and Alabama prepared to vote in union certification elections, Governor Brian Kemp cosigned a joint statement with five other Southern governors disparaging a United Auto Workers organizing drive as "special interests looking to come into our state and threaten our jobs and the values we live by."[151] A week later, he signed a bill restricting companies that agree to voluntary recognition of a labor unions through card check rather than require an NLRB election from accessing state tax credits.[152] A former Building Trades researcher summarized, "the union situation is tough here in Georgia, no doubt about it."[153] Despite these "tax incentives they [companies] get from the from the governor," a union leader explained that the unions are still getting work. In one battery plant, "the unskilled labor that they hired to put in that project screwed up so bad that they had to bring in unions to fix the problems."[154] While the Building Trades have aggressively courted contractors by promoting their skill premium, they have also worked in broad coalitions with racial and immigrant justice organizations to challenge the narrative promoted by the governor and argue that all jobs, and especially when incentivized with public funds, should be good jobs that respect workers' expertise, skills, and dignity and their voice on the job.[155] These coalitions represent a model for mobilizing workers and communities to support a just energy transition that works for the common good even as large questions remain such as how to protect incumbent workers, shift the model of contracting out solar by the utility, end the continued bifurcation of jobs in the clean energy sector, and expand organizing to credential, upskill, and represent nonunion workers to build worker power.

In practice, significant challenges remain in trying to realize a just transition for workers in the energy sector and across the economy more broadly where social reproductive labor remains unpaid or underpaid. Further, facing these headwinds at the state level and the Trump administration's efforts to roll back the IRA and other Biden-era policies, during the 2025 IRP, Building Trades unions advocated for continued fossil fuel use and new nuclear capacity, promoting energy *additions* rather than a transition.[156]

Unions are a mechanism long authorized and encouraged by federal law—if defanged and beleaguered by anti-union state policies—for workers to come together to fight for and secure the means of social reproduction for working people. Union members engage in collective action to improve the lives of workers on and off the job. As the production of electric life changes, unions have a key role to play to reject the hierarchical structuring of the labor market, the theft of time through unfair or discriminatory wages, inadequate benefits, and unsafe working conditions, and build worker power to advance a social wage.

Conclusion

The utility unions have been able to weather the hostile environment for workers in Georgia where pervasive anti-union laws, union busting, and the refusal of companies to bargain in good faith when their employees organize is commonplace, in part because of the relationships they have built with Georgia Power. The time that members spend in advocacy, organizing, and—as the next chapter discusses in greater detail—training the energy workforce all contributes to the social reproduction of electricity capital. Unions have been political allies in the ideology of growth (and, hence, job creation), supporting the utility and Commissioners in moments of political controversy where working class, environmentalist, and consumer advocacy organizations contested the profit-driven model of the utility. And yet, the utility maintains power over workers and benefits from a bifurcated employment model based on a constructed division between skilled and unskilled work. Unions in the clean energy supply chain have organized in response, signing up unorganized workers at jobsites where they should be receiving prevailing wage, building new alliances, and working to recruit and support marginalized workers.[157] The union is one site of labor organizing where power relationships based in racial and gender difference have both been reproduced and contested. Unionism can unify multiracial, working-class people across jobsites and creates a forum to organize democratically to demand the full value of their time. As one community-labor organizer described, a just transition would be: "Lower energy bills. More money in the pockets of the Southern worker. Healthier communities. Stronger democracy. A healthier and safer environment. A climate that is conducive to be able to support life for an extended period of time."[158] To organize the South is to work across difference and forge solidarity against the political and economic power of racial capitalism to demand the full benefits of a social wage including climate, energy, and environmental justice.[159]

5 The Work of Finding Work

In 2004, Van Jones launched the "Green Jobs, Not Jails" program with the Ella Baker Center for Human Rights. The Center was founded to honor Ella Baker's legacy of organizing throughout the long civil rights movement by bringing together Black, Brown, and poor people to build power to stop mass incarceration. Green Jobs, Not Jails confronted what Jones termed, "the two worst aspects of the present US reality: we have throw-away people, throw-away neighborhoods, throw-away children, thrown into prison, and we also act as if we have a throw-away planet."[1] Green jobs were defined as "family-supporting, career-track, vocational, or trade-level employment in environmentally-friendly fields," with most job creation expected in construction and manufacturing.[2] The campaign aligned with the Apollo Alliance, which later merged with the BlueGreen Alliance—both coalitions of unions and environmental organizations working to disrupt organized labor's opposition to climate policy by pushing for measures that would realize the job creation potential of clean energy.[3] In Oakland, the Ella Baker Center, Apollo Alliance, and IBEW partnered on the Green Jobs Corps, a green workforce training program.[4] In the mid-2000s, the Green Jobs Corps model gained traction and spread from California to cities across the county.[5]

Although green jobs united environmental, labor, and racial justice organizing, it faced limitations in implementation.[6] In an interview published in *Antipode*, Jones responded to criticism that his project worked within a neoliberal framework of entrepreneurialism in which marginalized workers could enhance their earning potential through skills training and "social and racial justice can be cast in terms of profit, value, and investment" in human capital. He dismissed the critique, retorting politics must offer a "practical program" and "coherent set of policy proposals."[7] In practice, the tides of neoliberal workforce development policy would work against the promise of the green jobs program as *job training* proved easier to deliver than *jobs*.

When the financial crisis rocked the global economy in 2008, international calls for a Green New Deal and national proposals for green Keynesian policy gravitated around green jobs.[8] Unions and environmental organizations pushed the Obama campaign and later administration to include funding for green infrastructure in a federal stimulus package, stating that work installing and manufacturing solar, energy efficiency, and other clean energy components provided "a short-term fix for the recession" while building "the future of our nation and our planet."[9] The Apollo Alliance estimated a carbon tax could spur five million new jobs that would reverse the loss of middle-class jobs through "mergers, bankruptcies, outsourcing, automation, and globalization."[10] Yet, the proposed cap-and-invest program to spur green job creation would later die in Congress after negotiation failed on the Waxman–Markey Bill.[11] Instead of comprehensive climate action, limited programs were included in stimulus funding embedded in the American Recovery and Reinvestment Act (ARRA) of 2009. Nearly eight billion dollars were allocated toward renewable energy and energy efficiency projects with a significant increase in weatherization assistance. The Green Energy Jobs Act, which Jones had championed "to train 30,000 young people a year in green trades," was also funded, launching training programs throughout the US, including in Atlanta.[12] President Obama appointed Jones to be green jobs czar, to work among federal agencies, cities, community groups, and unions. Despite broad coalitional support to create pathways into good jobs in weatherization and other clean energy projects,[13] long-term employment failed to materialize. Without a carbon tax to incentivize climate action, there was limited new labor demand. Instead, the supply-side emphasis on job training created a pool of workers for whom there were no jobs after federal resources dried up. As a result, green jobs have been largely disparaged by workforce development professionals and abandoned by the federal government, which stopped tracking them in 2013.[14]

The nebulous vision of green jobs was out of step with actual employment opportunities at Georgia Power or other energy employers. In an interview, Laura, a program officer with a workforce intermediary located in the Atlanta United Way that recruits and connects workers to training, recalled: "For a while, green jobs was sort of like the flavor of the day. . . . Workforce people go, phew, when they remember it," she sighed as she shook her head. There was "this *huge* Department of Labor funding . . . for 'shovel-ready jobs,' and green jobs was the big thing they threw all this money at." But it was not successful, she continued, because green jobs "just doesn't mean anything in our world."[15] For training providers, green jobs, or even the notion of greening the energy sector, was not specific enough to support workers.

The promise of job creation in clean energy continued to animate progressive calls for a Green New Deal and were embedded in the IRA where a revised focus on registered apprenticeship as an on-the-job training model in specific trades responds to the prior decade of green jobs failures.[16] In practice, all kinds of jobs need to become "greener" to address climate change.[17]

Despite the perceived failures of these programs, training for green jobs remains a seductive and bipartisan rallying point for environmentalists, industry, and public officials. Atlanta's 100 percent clean energy initiative promised to create eight thousand jobs and outlined the need for "coordinated efforts on workforce development between relevant City offices and departments, technical training programs, and employers [that] could help grow a sustainable market for these services, producing long-term, stable employment and opportunities for new businesses and growth."[18] The Office of Sustainability, which wrote the plan, imagined that if public resources could create a pool of trained workers, a market for their services and stable employment would follow. These calls for green jobs training—often produced by highly educated academics and researchers biased by their own educational trajectories—began from a flawed assumption that labor supply, not labor demand, is the issue. Building on neoliberal human capital theory, training-focused strategies assert skilled workers will be more competitive in a labor market. In the process, these strategies implicitly blame individual workers and their perceived lack of skills for unemployment or underemployment. Although real training and education needs are a prerequisite to getting hired in many jobs, in *The Job Training Charade*, Gordon Lafer aptly diagnoses the flaws of this logic that assumes plenty of good jobs are available if only people have the determination, perseverance, and skills to find them. In practice, as this book has so far demonstrated, the failure to address climate change and create clean energy jobs has little to do with the skill deficiencies of workers. More robust training programs cannot conjure a market for green employment or improve the quality of jobs so that workers might want to take them.

The rhetorical promise of green jobs fails in two ways. First, there is a disconnect between green job *creation* and job *training* that takes place in and beyond the electric utility. Second, training is inadequate to address poverty and overcome racial and gender discrimination in employment, despite being a preferred workforce strategy to manage the low-wage labor market.

This emphasis on job training is relatively recent. By training, I refer to a broad category of classroom and on-the-job instruction designed to prepare or upskill workers for employment. Two- and four-year colleges and universities, technical schools, union and nonunion apprenticeships, employer programs, and nonprofit entities all provide training, molding a surplus reserve of most often young and un(der)employed people to the temporalities of wage labor. The public workforce system and local workforce boards funded through the federal Workforce Innovation and Opportunity Act (WIOA) also oversee workforce development programs for marginalized and dislocated workers. The creation of local workforce boards accompanied "workfare" reforms in the 1990s that dismantled universal (if limited) welfare programs and made receipt of social services for many contingent on work or work activities. This shift accelerated the rise of short-term, work-readiness programs at the lower-wage end of the labor market. Job seekers enrolled in classroom and on-the-job training are eligible for career coaching and wraparound services such as financial assistance for food, transportation, housing, childcare, relocation expenses, and legal services. Workfare rescaled (albeit limited) collective institutions of social welfare by individualizing responsibility for social reproduction and reorganizing social services as recruitment and training opportunities for employers.[19]

In this chapter, I historicize the discourse of green jobs that dominated discussions of clean energy nationally and in Atlanta in the 2000s, where multiple green jobs programs started. Reviewing the limitations of these programs, I consider how training functions in clean energy policy to both create and manage a workforce for the electric utility and related energy industries. Following Jamie Peck's etymology of workfare, I trace the role of training through the shift from the New Deal to the neoliberal order defined in part by the turn to workfare that reorganized state capacities to address poverty and shifted responsibility from the state to individuals and families.[20] I contrast two green jobs programs, the first for non-profit weatherization assistance and the second through the Technical College System of Georgia. The latter has been promoted and used by Georgia Power.

Although I critique the uses of training as a supply-side intervention to cheapen labor, I do not debate the value of education and the need for vocational training to support workers' placement, advancement, and fulfillment. Instead, in asking how training is delivered, to what end, and why the benefits of social reproductive services remain focused on job seekers, I argue that training is an insufficient measure to address poverty and low-wage work. Instead of pushing onto workers "the work of finding work," advocates of green jobs can embrace a wider vision of energy democracy based on building a social wage such that social reproduction is ensured for all regardless of one's paycheck.[21]

The Birth of Training

Social reproduction and vocational training have a long and intertwined history in Atlanta that begins with fragmented local efforts to manage poverty, including programs led by Black civic leaders.[22] During Reconstruction, universities now associated with the Atlanta University Center were established, which made Atlanta an "unparalleled center of African American higher education" and forged "its later reputation as a 'Black Mecca.'"[23] Teachers, social workers, and bureaucrats also founded social institutions committed to vocational education as a way to "improve" social and political culture and advance racial uplift of poor and working-class people to win the full benefits of citizenship for Black Atlantans.[24] Among the most well-known educational programs in the New South was Booker T. Washington's Tuskegee Institute, which became the "symbol of racial progress" where "formal training in the values and skills necessary for bourgeois success—was viewed as the beginning of mass liberation."[25] The vision he charted for "the educational needs of black [worker]s in terms of proper training, not state financial support," harmonized with white Southern elite who viewed social reproduction as the responsibilities not of the state but the family, church, or community.[26] In Georgia, for instance, the state constitution barred public spending on most poor relief that could undercut the low-wage economy.[27]

In 1895, members of Atlanta's urban regime hosted the Atlanta Cotton States and International Exposition, which sought to showcase the advances in Southern industrialization and racial reconciliation. Exhibitions chaired by prominent citizens included the Electricity Hall designed under the leadership of the future chairman of the Georgia Power board, H. M. Atkinson, and the Negro Building highlighting African American craftsmanship and art, chaired by Washington. On the first day of the exposition, Washington delivered a speech to a crowd of white and Black on-lookers, proclaiming in a now infamous speech that "in all things that are purely social we can be as separate as the fingers, yet one as the hand in all things essential to mutual progress."[28] His address legitimated segregationist policies that undergirded the progressive politics of the New South and was reprinted with praise in the *Atlanta Constitution*.[29] Du Bois criticized Washington for accepting incremental steps toward civil and political rights for African Americans and rebuked Washington's "unnecessarily narrow" industrial education program that urged Black people to "concentrate all their energies on industrial education, and accumulation of wealth, and the conciliation of the South."[30] In particular, he disputed Washington's assertion that Black people's "future rise depends primarily on his [sic] own efforts," which "has tended to make the whites, North and South, shift the burden of the Negro problem to the Negro's shoulders . . . when in fact the burden belongs to the nation."[31] In other words, the poverty of Black workers had little to do with their personal failings or lack of skill or preparation to succeed in industry. Education could only do so much to remedy the failure of the nation to provide reparations for the theft of unpaid labor from formerly enslaved workers and racism reinforced through legal, political, and informal means that limited access to jobs, property, and wealth accumulation.

Despite these critiques, during the Great Depression, educational and vocational training received new attention as part of New Deal programs to provide welfare, housing, and employment with the federal government hiring unemployed workers in public works.[32] Initially, these programs were administered through a segregated state bureaucracy. Governor Eugene Talmadge, who took office in 1933, sought to control federal funding through personal networks of patronage that distributed emergency and work relief to county-level committees staffed by agricultural and business leaders. The Fulton County relief committee was chaired by Georgia Power general manager and board chair W. E. Mitchell, who was not shy about disclosing his personal distaste for the poor, saying: "I don't want to see them, don't let me see them. There will be causalities of this movement [the Depression] and we must steel our hearts against them all as we did against wounds and death in the war."[33] In contrast to the genial commitment to public service embedded in the company's slogan, "A citizen wherever we serve," coined just a few years prior, Mitchell denigrated the welfare programs he was set to administer that could challenge the racialized, low-wage labor market on which he and other middle-class people relied to staff their homes and businesses.

Recognizing state intransigence, the federal government returned New Deal programs in Atlanta to federally appointed administrators.[34] In contrast to the state government that ignored or actively weaponized state power against African Americans, Karen Ferguson describes these administrators, who were often white women social workers, as more sympathetic, yet patronizing. Implementation of relief efforts did not disturb "the South's vocational apartheid."[35] Work relief programs intended to employ out-of-work adults and youth were segregated. White boys and men were afforded opportunities to train and work in skilled construction; Black boys and men were more often relegated to unskilled industrial and agricultural labor, while programs for white women encouraged domestic science for future homemakers and trained Black women as domestic workers. Black activists and social workers who staffed the Colored Division of Atlanta's National Youth Administration resisted the use of training to segment the labor market and introduced vocational training in skilled labor that white unions and industries refused to provide.[36] Black tradespeople and union members also organized and petitioned federal administrators to secure quotas established for Black employment on public works.[37] Training was only a small piece of wider New Deal social welfare policy and coupled with public employment, yet training would grow in importance as a strategy to support workers in periods of economic transition. Most notably, the subsequent adoption of the GI Bill after World War II subsidized education for veterans to address unemployment and ease the shift from wartime production.[38] However, in each instance, training served not to equalize opportunity but reproduce the racialized and gendered division of labor.

The roots of contemporary training infrastructure stem from the 1962 federal Manpower Development and Training Administration, which initiated both stipends for classroom training and on-the-job training in which employers received a subsidy to hire workers displaced by automated and computerized production.[39] On-the-job training opportunities for skilled workers remained targeted to white men. Further, in Atlanta the Georgia state employment agency made many more programs available to the white DeKalb Technical and Vocational School than the Black Carver Vocational School, justifying that providing the complete curriculum to Black schools would be "costly and useless," given craft, technical, and white-collar positions were not open to Black workers.[40] Black and feminist activists questioned how workers without access to training could advance if openings were available and called out the failure of limited classroom programs to address high unemployment among Black men, young people, and women.[41] For instance, when the *Atlanta Constitution* surveyed employers to assess whether they actually valued credentials from federally subsidized and private courses for key punching intended to train women as clerical workers, most, from Coca-Cola to regional banks, explained experience was more important, with only Georgia Power even giving a credential consideration when reviewing applications.[42] Georgia Power not only committed to hire graduates of public vocational programs but also sponsored programs such as appliance-repair courses. "When industry helps educators plan training courses as we have done," explained Georgia Power Industrial Relations manager W. Huber Joiner, "it can then have a chance to be as delighted with the results obtained in the young men trained in these courses as we are."[43]

While racial justice and feminist movements fought to desegregate apprenticeships and civil service exams as roadblocks erected to bar Black workers, workers of color, and women, many also recognized training was insufficient to address labor market segmentation and advocated forcefully for public employment to provide work for poor and unemployed people.[44] They argued public employment could provide a wage floor and mitigate structural and regional unemployment due to energy-intensive mechanization and capital flight. Such policies faced opposition from Southern Democrats, who argued that the role of government was not to create jobs but to facilitate the private sector. During the economic crises of the 1970s, limited public employment was dismantled in favor of public subsidy for training as the means of reproducing workers, epitomizing an ideological shift away from the New Deal order as neoliberal workfare policy ascended. This shift subverted early and on-going efforts to create green jobs.

Green Jobs to Green Jobs Training

The notion of "green jobs" first arose from the conjuncture of civil rights activism, growing environmentalism, and the 1973 energy crisis. Tracing the evolution of training in Atlanta reveals the ways in which the public promise of employment intended to support workers in securing the means of reproduction was refashioned through multiscaled public policy as city officials faced austerity and Georgia congressmen advocated for workfare. The neoliberal reforms that put lifetime limits on public benefits and added work requirements to welfare expanded the Southern imagination of the state as a tool to promote industry in recalibrating limited social support for nonworking people to provide training and cheapen labor for employers. This section is not an exhaustive history of welfare reform's effects on the labor market[45]; instead, I detail the conditions in which green jobs policy emerged and began to spark before catching fire in the financial crisis.

In the 1970s, the federal government was looking for answers to address falling productivity, persistent inflation, and unemployment. In 1975, Senator Hubert Humphrey took the Joint Economic Committee of Congress on a research mission to hear from labor leaders, community organizers, and public officials at field hearings in Atlanta, Boston, Chicago, Los Angeles, and Fall River, Massachusetts.[46] In Los Angeles, Derek Shearer, codirector of the California Public Policy Center, argued that the "protection and renewal of the urban and rural environment can be a source of millions of jobs."[47] He submitted reports calling for a "green collar revolution."[48] This first mention of green-collar jobs took inspiration from New Deal employment programs like the Civilian Conservation Corps,[49] suggesting a federal green job guarantee that went beyond the ambitions of on-the-job and classroom training programs and sought to "substantially alter the structure of the labor market" that "segregated minorities and women into certain kinds of dead-end jobs."[50] With the federal government as an employer of last resort, workers would have more choices and private employers would have to improve wages and working conditions.

In the Atlanta hearing, Mayor Maynard Jackson also endorsed a full employment policy, greater funding for cities, and a permanent extension of the Comprehensive Employment and Training Act of 1973, which provided both vocational training and funding for public employment. He concluded in saying: "I tell you now my number one frustration as mayor is to see people not begging for welfare, although they need it; not begging for food, although they are hungry; not begging for money, although they are broke; but begging for a job."[51] The unemployment rate in Atlanta had risen from 1.7 percent in 1971 to 12.5 percent in 1975 and 15.3 percent for Black workers. Nearly 20 percent of people in the city were being paid through federal programs, indicating that the crisis could not be managed with the resources of the city alone. Clark College President Vivian Henderson also pointed to the possibility of green jobs, observing the successful integration of construction work for water infrastructure that employed Black youth following community and labor union organizing to pressure the Army Corps of Engineers to implement local hiring goals for rural development projects in south Georgia. He urged this example could be replicated to ensure equal opportunity "for all races, minorities, and sexes" in public works.[52]

Public employment was popular across racial lines and strove to keep the New Deal coalition of labor unions and people of color united behind the Democratic Party while incorporating the objectives of a growing environmental movement. Shearer explained in an interview that the full employment policy united these factions because "it doesn't matter what color you are, you need a good job, you need affordable housing, you need health care. And so that should be the glue—the economic interest. . . . The thinking would seem to me, it's pretty common sense, politically."[53] Yet, as he went on to explain, "Starting with Nixon . . . [the Republican Party] had a southern strategy to use race as a political weapon . . . to turn the South Republican. . . . That right-wing resurgence [prevented] doing a lot of these things."[54]

Ultimately, green Keynesian policies were not taken up. A full employment bill proposed by Senator Humphrey and Representative Hawkins was stripped away until only a commitment to balanced growth remained. This attempt "to revive postwar liberalism became the New Deal that never happened," mourned historian Jefferson Cowie. The failed legislative push ended as "something closer to a requiem for a collective economic vision for the American people."[55]

Instead of a comprehensive strategy to suture the leaking social safety net in the US, Southern Democrats sought to dismantle New Deal welfare protections that laid the foundation for the 1990s overhaul of welfare. Georgia Senator Herman Talmadge, son of Governor Eugene Talmadge, proposed the Work Incentive Program that was adopted in 1971 and required welfare recipients whose children were in school to work.[56] The subsequent Job Training Partnership Act of 1982 eviscerated public employment and revamped short-term training initiatives to encourage work readiness.[57] President Reagan justified this move away from public employment programs and to training "by insisting that there already were enough jobs for those prepared to take them."[58] In an extensive review of the legislation, Lafer diagnoses that workers' preparation was not the barrier to full employment. A survey of employers found

> by far the most common complaint of employers focuses not on any traditionally defined skill at all, but rather on workers' discipline, punctuality, loyalty and "work ethic." However, it is unclear that punctuality and discipline can accurately be termed "skills": they may constitute acquired habits, but they are also in large part commitments which one chooses to offer or withhold on the basis of the wages and conditions provided. To the extent that employers are simply paying too little to get the work commitment they desire, this is a mismatch of a different sort.[59]

Reagan-era reforms focused on the supply side of the workforce, trying to correct workers rather than improve the quality of jobs. Despite the shortcomings of this approach, a discursive focus on training or workforce development remains in federal funding for work readiness that was updated in 1998 and again in 2014 as the Workforce Innovation and Opportunity Act (WIOA).

Teaching potential future workers "soft skills" such as timeliness, work ethic, and courtesy remains a core part of Atlanta's various WIOA-funded training curricula. In 2019, I visited Atlanta Builders,[60] an intensive "boot-camp" style construction training program. The program started before construction began on the Atlanta Falcons Mercedes-Benz Stadium to recruit workers from the Vine City neighborhood and service contractors who were struggling to meet the city's minimum threshold for minority employment on projects receiving public funding. Atlanta Builders took on the labor of vetting and training people for employers and has become one of the largest construction training providers in Georgia. In some ways, Atlanta Builders embeds the principles of the Green Jobs, Not Jails program by working to connect WIOA-eligible residents living in disinvested neighborhoods to construction employers and teaching OSHA safety protocols, first aid, construction math, construction drawings, tool identification, forklift safety, and utility flagger certification, as well as "soft skills" in communications, employability, professionalism, interviewing, resume building, and financial literacy. The program meets five days a week for nine hours a day over a month. Sally, a former graduate of the program who had come back to support recruitment, described, "I've had people say 'Twenty days? how you teach somebody construction in twenty days?' No, we're not teaching them construction. We're teaching them how to work in the industry and how to advance."[61] Heather, a program administrator, added that soft skills are job-survival strategies in the construction industry where "bubbas" can be hostile to Black and nonmasculine people. She shared some of her tips. Staying in a job is sometimes just knowing how to look busy: "Be busy the whole time. Do not ever be slack, on a jobsite, always have something to do, sweep the floor, pick up cigarette butts. Just put your ears on and keep your mouth shut."[62] Trey, a technical college administrator who helped to start one of the first green job training programs in Atlanta, added in an interview that training begins in bringing people up to speed academically and addressing "the lack of soft skills. So, we can train them in a classroom all day long. . . . But if Jana doesn't understand that you got to report to work on time, you got to dress a certain way, you can't do drugs. You know?" he paused, looking at me, letting it sink in for emphasis before continuing with much unsaid. "So, we have to also take time to ensure that level of education is disseminated."[63] Soft skills training is code for respectability politics that discipline workers to the temporalities of waged work and reinforce, both materially and ideologically, that work is necessary for livelihood generation. These worker-focused interventions do little to require employers to provide jobs worthy of people's time and attention or address discrimination trainees and workers face.[64] Instead, training programs reproduce worker subjectivities, which like "every relationship of 'hegemony' is necessarily an educational relationship."[65]

Training flourished within a workfare regime espoused by Atlanta-area Congressman and House Speaker Newt Gingrich, who asserted the need to restore personal and family responsibility through marriage, work, or training to prepare for wage labor.[66] The 1996 Personal Responsibility and Work Opportunity Reconciliation Act championed by Gingrich accomplished this "tectonic shift" to enforce work discipline and reduce welfare expenditure, transforming the already-inadequate New Deal–era Assistance to Families with Dependent Children welfare program into the time-limited Temporary Assistance for Needy Families.[67] The changes de-emphasized cash assistance to give new priority to social services that promote work readiness. Corresponding cuts in higher education, affordable housing, and healthcare pushed provisioning for social reproduction from a public responsibility onto individuals and family networks, leading to a rise in both debt financing and charities seeking to close the gap in public service.[68] In Georgia, immediately after the reforms, the total share spent on low-income services declined, even as spending for work-related services increased with a range of customized wraparound services made available to job seekers including childcare, transportation, substance abuse treatment, and career counseling.[69]

Public and nonprofit training providers in Atlanta continue to receive bipartisan support. Jerry, a deputy commissioner with the Technical College System of Georgia, explained in an interview that for students who are eligible for WIOA based on their income or zip code, "We'll help you with the tuition [for technical college], we'll help you with any of the costs and if you have children, help with childcare while you're in school. It's kind of a blanket to help people get off of welfare or never get on welfare. That's really the way you can look at it. So, I joke that this is one of the few federal programs that all the political spectrums are okay with. You know what I mean? I mean, nobody argues with this stuff."[70] At Atlanta Builders, job seekers can receive food assistance, a MARTA card for the city bus and train system, and legal assistance to apply for a driver's license, license reinstatement, or criminal expungement. Services support participants through the training program, "so they're not worrying to death about what their next step is," explained Heather.[71] Within the anemic remains of the New Deal social welfare state, training programs are a politically neutral way to manage low-wage labor markets, paying subsistence stipends for people out of work when there are not sufficient jobs available, offering lessons in "high-demand" careers tailored to employer needs, and providing vital social services for participants engaged in training to become a worker.

Workfare cannibalized the biopolitical capacities of the welfare state to protect and shepherd life and replaced it with new bureaucracy to protect and expand the labor market. The deeply unequal reverberations of the collapsing social wage—and the rise of policing and mass incarceration—left criminalized, ghettoized, and "deviant" communities to provide for themselves amid the withdraw of public support to regulate labor and social reproduction.[72] Workfare programs nonetheless draw on the individual, familial, and cooperative care people undertake to make life amid organized abandonment to cultivate an available workforce.[73] Training is one tool through which the state guides "human capital" investment, as Wendy Brown describes, pushing workers to improve their "competitive position and appreciate in value" by cultivating professionalism, responsibility, and skills.[74] Participation in training is often coerced to satisfy workfare, probation, or parole requirements that compel individuals to take low-wage jobs and put a ceiling on their opportunities for advancement.[75] This takes shape in the form of "benefit cliffs" for food stamps, Medicaid, or Section 8. As the salary of someone goes up, Laura described

> Public assistance doesn't go down proportionally. It stays with them until they're no longer eligible, and then it's gone, so if you look at the effective wage rate, which is what they're earning as well as the assistance . . . they can end up so much worse than they ever were. . . . You will have individuals who will turn down raises or promotions, because effectively they're going to earn less and you can't blame them for that because they're barely making ends meet. . . . We're all trying to move nursing assistants up to medical assistants and patient care techs because we think those are better jobs and they are in the hourly rate, but they're still not going to get someone to a sustainable wage job because as soon as they lose their assistance, they're back down at the low wages.[76]

The result is that training initially intended as an educational pathway to improve the skills and, hence, wages of an individual intersects with low wages, anti-poor and anti-union regulations, discrimination by employers and social service providers, and a general lack of social reproductive services to reinforce the segmentation of the labor market.[77] This traps low-wage workers in a narrow range of paid work in which they are not making enough to support themselves or their families without social assistance, but risk losing assistance if they move into a job making more. Alexander Camardelle argues WIOA-funded programs for "opportunity youth" in Atlanta maintain the segmentation of the labor market and "advance racism in their everyday, state-led objectives to create a talent pipeline for employers" that "fail to serve as a pathway to economic liberation for Black youth."[78]

The revival of green jobs in the 2000s challenged the neoliberal consensus around workfare and mass incarceration as bipartisan strategies to manage surplus labor. Outside of scattered mentions of the green employment in conservation and environmental fields,[79] green jobs largely faded from view during the first Bush administration, with the notable exception of Al Gore's insistence on the 1992 campaign trail that there were "millions of green jobs to be created" in addressing climate change. President Bill Clinton affirmed this opportunity in his Presidential Earth Day address in 1993.[80] The following year, he signed Executive Order 12898 on Federal Actions to Address Environmental Justice in Minority Populations and Low-Income Populations in response to organizing for environmental protection.[81] Environmental justice leaders, including Robert Bullard, made the case that environmental justice is a labor issue because polluters use "jobs blackmail," forcing workers of color who are already constrained to "low-paying, low-skill, high-risk blue collar and service occupations" to take positions that they know are bad for their health because it is the only job available.[82] Envisioning a different labor model for environmental good, the Green Jobs, Not Jails campaign took inspiration from environmental justice, labor, and civil rights organizing, gaining national attention amid the financial crisis. Green jobs programs that took root in Atlanta require greater scrutiny to understand their limits to subvert workfare pressures focused on training.

Weatherizing the Recession

The recession sparked myriad programs in Atlanta to train a new generation of workers using the influx of federal funding for energy efficiency. In 2008, the state set out targets for energy savings in public buildings, and in 2011, the City of Atlanta signed onto the Better Buildings Challenge to reduce energy and water consumption in commercial buildings 20 percent by 2020 in partnership with Central Atlanta Progress, the Atlanta Downtown Improvement District, Midtown Alliance, Livable Buckhead, and Southface.[83] At a national forum on green jobs hosted by the Emerald Cities Collaborative, Deborah Scott, from Georgia STAND-UP, couched the energy efficiency measures as a historic opportunity to address persistent poverty in Atlanta that remained despite the success of the biracial entrepreneurial class of the urban regime. She noted, "Thanks to Maynard Jackson, we have a very aggressive minority [contracting] program where we've made a whole class of black millionaires so our [issue] is more of a class thing. So, it's race and class. So, we have Black folks that are doing very, very well, but then we have the haves and the have-not—if unemployment is 15% in Atlanta, black unemployment is 30%, I mean people have just fallen off the grid."[84] The grid here is a metaphor, to describe the labor reserve, who green jobs policies targeted. Optimistically, Scott asked how the mayor's office would reverse the city's historic acceptance of racialized poverty to realize Van Jones's vision of un(der)employed Black people becoming taxpayers and employees doing meaningful work in their own communities while improving neighborhood and public infrastructure.

Despite initial support from community organizations, labor, and businesses for the Better Buildings Challenge, Rebecca, an energy advocate and former participant in the Green Jobs Task Force formed in Mayor Kasim Reed's administration, recalled problems emerged in moving green job creation forward. She said, "It came under Obama, and you have Van Jones representing it. And it was *the* sexy thing to do, right? And Atlanta being a majority minority city, and having some labor union presence, there was a desire by different groups to do something on green jobs, to figure it out. . . . And I think it's suffered from things we still see green jobs suffer from today is like, what do you mean? What are you actually going to do, more than just talk about it? And so we just kind of floundered."[85] As noted in chapter 4, interest from labor waned. The city spent its stimulus funding to design its Climate Action Plan, hire staff, complete energy audits of municipal buildings, and implement projects that would have the highest return on investment.[86] While federal money financed various programs, the counterhegemonic ambition to defy the low-wage labor model, or even reach the green job growth charted in city sustainability plans, was hard to attain. In practice, energy efficiency implementation fragmented into numerous training programs, with unions getting some of the work and several new training initiatives starting for weatherization. While Georgia Power offered some funding to top up federal weatherization assistance provided by nonprofits, at the time it remained reticent to act on anything green.

The most significant effort through the environmental nonprofit Southface was the Southeast Weatherization and Energy Efficiency Training (SWEET) Center. More than 1,200 people received training in residential retrofitting and energy efficiency to work with community action agencies, weatherizing several thousand homes during the training programs.[87] In conjunction with AmeriCorps and YouthBuild, Southface also trained young people in sustainable building principles and energy efficiency audits.[88] Each YouthBuild cohort constructs at least one affordable housing unit, which is targeted for sale to low-income, first-time homebuyers. Trey explained that incorporating energy auditor training seemed like a natural fit to introduce participants to green career fields, "It was a great opportunity for the kids." He continued, "We saw a potential market for this work as an opportunity to go out and try to build a program that we could go into the community, especially with elderly [people] and do this [energy efficiency improvement] work and unfortunately, that didn't take off like we thought it was going to."[89] While Southface had training programs to "upskill" contractors and builders that are still running at the time of this writing, Trey reported, "There was not really a lot of new opportunities to train in this energy efficiency space at least for young people." The hope of creating "a new generation of professional technicians" simply didn't happen, because the demand was not there for energy efficiency services.[90] Without on-going government or ratepayer funding for energy efficiency to create demand for workers, the promise of green jobs eroded. All that remained was the training program without a pathway to employment for the graduates.[91] Eventually, Southface stopped doing weatherization training altogether because they could not place graduates into jobs.[92] Several years later, YouthBuild donated the equipment it had purchased with stimulus grants to the only student from the program who was still in that line of work.[93] Despite the promise of future employment in Atlanta's 100 percent clean energy plan, Josh, a workforce development coordinator, said Southface would not revive training for green jobs because they do not want to "train to unemployment."[94]

In opposition to the neoliberal philosophy that vocational training and higher education are an investment in "human capital," green jobs training often failed to appreciate the value of a worker's time.[95] If the state had set energy efficiency savings targets or Georgia Power had authorized funding for energy efficiency, the pathway from training program to career might have been formalized as advocates intended. The young people in YouthBuild and other graduates of the SWEET Center could have moved into jobs, advanced through skill acquisition to being apprentices, started their own contracting businesses, become journey-level workers in a trade, or worked up the career ladder at a utility. Without connecting the supply- and demand-side pieces, green job programs were little more than training that put people to work preparing for work using the confluence of stimulus funding and social service programs to keep workers "on the grid." A federal government employee who had been in charge of administering stimulus funds locally admitted there was little accountability to the vision of creating career pathways in green jobs. Green was "a little icing on the cake. . . . But when we report to Congress, we're not reporting green, we're not reporting workforce, we're reporting 200 jobs, $70 million of private investment, and Congress doesn't care necessarily what those jobs or private investment is doing. . . . We're in the public infrastructure business. We're not in, you know, policing companies and how much they pay their employees."[96] Federal policy ambitions to address racialized urban inequality, climate change, and energy burden failed in practice to change the intransigent regional development agenda.

Despite the frustration of green jobs programs to do any more than train workers, federal industrial policy to reshore manufacturing of the components of electric life have revived discussion of green jobs in solar, battery, and electric vehicle manufacturing facilities and clean energy deployment. Following the passage of IRA, new industries faced stiff competition for a limited pool of workers with record low unemployment and declining labor force participation as the neoliberal disinvestment from social reproductive services means many potential workers are too sick, immobile, or busy with child- and eldercare to participate in the labor market. Facing constraints in the supply of available workers, the state of Georgia and Georgia Power have turned to training. Training—more transparently than public education—employs state resources to capture the time of potential future workers to recruit industry and advance the state's economic development agenda.

Quick Start on the Battery Belt

Amid the failures of green jobs programs that worked through the city and nonprofits, technical college programs were able to weather the recession and create pathways to employment.[97] When I asked Jerry from the Technical Colleges about green jobs, he remembered the stimulus for the emphasis it gave to solar manufacturing, explaining, "There was this big, 'we're going to go solar' . . . [so] we got with industry and worked with that. A lot of that industry didn't make it, but some . . . are still in operation and going strong, and we're still providing them with a workforce." The technical colleges partnered with Suniva, which opened a manufacturing facility in Dekalb County in 2008.[98] Jerry described that clean energy manufacturing was not a bust, like in the residential energy efficiency sector, but at the time, it was "just not the explosive thing that they thought it was gonna be."[99] Explosive industries are what the state training infrastructure is intended to serve. The twenty-two technical colleges across the state work directly with industry to develop targeted training programs in areas with a clear growth trajectory. Training people in high-demand careers helps them find a job at the end and also meets the needs of employers making new investments in the state by recruiting potential employees.

Since 1967, the technical colleges have offered the Quick Start program. This state-funded initiative was separate from the federal Manpower Development Training Program and had the specific goal to recruit employers to

Georgia and "to prepare a workforce accustomed to seasonally dependent, sunup-to-sundown farm labor to industry's regimented 8-hour workdays."[100] Through Quick Start, the technical colleges can provide a workforce assessment for a company investing or expanding in Georgia that will create at least fifteen jobs, create a customized training plan, and train the first cohort of employees without any charge to the company.[101] Where expansion of an industry is anticipated, the curriculum is turned over to the technical colleges so that "we can provide the pipeline . . . so there's always new graduates coming in[to employment] as people get promoted or transfer."[102] Many of these programs are highly specific—workers who are hired onto a job go through weeks-long to months-long modules on a specific software or part of an assembly line. Of the various kinds of credit, noncredit, and short-term training that technical colleges offer, Quick Start, continued Jerry, is "kind of one of our big secrets in Georgia about how we attract and do such good business for business."[103] The Quick Start program takes on training for potential employers, shifting the responsibility for reproducing the workforce from the company to the public and individual. Quick Start is another service alongside those offered for infrastructure development and site location that Georgia Power and county and state officials promote to manufacturers.[104]

The benefits for workers are immediate job placement. Trey argued that "the Technical College System of Georgia understands that everybody, first of all, don't have the academic ability to do this credit side [of an associate's degree] and also don't have the time. You need some quick hits, that's going to equip people with industry recognized credentials, that can lead to employment that's paying a living wage."[105] The technical colleges prioritize building pathways into jobs worth staying in that pay $20 or more an hour, where, in Trey's words, workers are "compensated in a way that would allow them to live."[106] In Atlanta, the reproductive justice living wage for a single-parent household to support a family was $28 an hour, well above the state minimum wage.[107]

Some of the most recent beneficiaries of the Technical College expertise are EV manufacturers. According to Georgia Power, between 2018 and 2024, more than forty EV-related projects have been announced, which are expected to total $27.5 billion in investment and create more than thirty-two thousand jobs.[108] These EV, solar, and battery plant location decisions build on decades of cooperation between utility, state, and technical college partners. In the 1980s, international automakers began to open US factories outside of the traditional stronghold of auto manufacturing in the Midwest and locate in the South given the (nonunion) wage levels, worker training, and incentive packages provided by state governments. As Timothy Minchin chronicles, Georgia was the only Southern state without an international automaker when Governor Sonny Perdue proactively lobbied to bring Kia to west Georgia. He appointed Craig Lesser, former vice president of government and regulatory affairs for Georgia Power Company, to run the Georgia Department of Economic Development in 2004 and negotiate with Kia.[109] The plant was described as "a godsend" to a former textile-producing region with double-digit unemployment when the facility opened during the financial crisis.[110]

Part of the package that lured Kia to Georgia was Quick Start. The technical colleges took on the role of hiring the first thousand employees from more than thirty thousand applications, matching candidates with different occupations. Kia Georgia's director of people and culture praised Quick Start for creating "a world-class training program to prepare the community for the 3,000 automotive assembly jobs headed its way."[111] *Atlanta Journal-Constitution* staff, who initially reacted to the $410 million in incentives given to Kia with skepticism, praised the significant ripple effects as the plant attracted suppliers that are now colocated in the region and additional revenue circulates throughout the state as the factory uses infrastructures from Savannah's port, Atlanta's airport, railroads, and Georgia Power's electric grid.[112] In 2023, Kia announced plans to add a new electric vehicle line, creating another two hundred jobs.[113]

Georgia's manufacturing sector has grown exponentially with state investment of taxpayer funding to line up social reproductive services, vocational education, and training to recruit and prepare workers for the clean energy transition. "This green wave wouldn't be possible without Quick Start," explained Georgia Department of Economic Development Commissioner Pat Wilson in an interview in *Site Selection* magazine.[114] An SK Battery manufacturing plant that broke ground in 2019 promised $1.67 billion in investment in return for $300 million in public funding. The Company's director of external affairs emphasized, "The key incentive for SK Battery is the Georgia Quick Start program. They know exactly how to train the average person off the street."[115] For solar panel manufacturer Qcells, Quick Start training materials were also "a game-changer" that cemented their decision to invest and expand in the state. A human resources director at the company shared that the state taking on training meant that Qcells "didn't have to invest the thousands of hours of making classroom and video training that was necessary" to stand up its $2.5 billion solar panel facility.[116] The seventy-five Quick Start employees, unencumbered by the strings attached to federal training funds, take on this labor of reproducing the workforce for companies.[117] Numerous battery component producers and recyclers have followed, motivated by public resources, workforce, and access to cheap and reliable electricity.[118]

The most significant clean energy beneficiaries of state programs have been two massive EV manufacturers recruited by state agencies and Georgia Power: a Hyundai plant in Savannah and the electric truck manufacturer Rivian southeast of Atlanta. In total, the state provided $3.3 billion in state investment to these two projects.[119] This included tax credits, bonds, land, and the full use of the Quick Start program, including building and operating a training center and recruitment program for Hyundai valued at upward of $140 million and another valued at $89 million for Rivian, including a commitment to designing up to forty-three different kinds of modules to train company employees.[120] In 2023, Rivian announced it would add on-the-job training, a priority of the Biden administration, with a twenty-four-month apprenticeship program for its employees that nonetheless continues to rely on public resources as "tuition will be covered by the State of Georgia's HOPE Grants."[121] In return for state funding, Rivian has agreed to invest at least $5 billion and create 7,500 jobs paying an average of $56,000 a year. The fine print of the agreement explains that the job-creation requirement will be fulfilled if Rivian creates or contracts the required number net new full-time jobs paying at least $20 hour for a minimum of thirty-five hours per week with the opportunity for access to, but not necessarily paid or subsidized, medical benefits.[122] The aggregated average wage masks an unequal labor force as the state allows lower wages and outsourcing without a guarantee of medical benefits.[123] Neighbors adjacent to the massive facility also sued to halt the project over environmental nuisances from construction and objections to the incentive package that provided property tax savings to Rivian, without addressing social reproductive concerns regarding schools, traffic, and housing for potential workers.[124]

The commitments of the county and state governments, the technical colleges, and the privatized proxy of the state, Georgia Power, made to recruit EV manufacturers to Georgia rely on trainees' social-reproductive labor to power the energy transition and sustain electric life. State training programs enlist the time of potential future workers as a mass of "nonsubjectified labor, or disposable life-times" to bargain with potential investors.[125] Georgia asserts that the technical college staff can find and train surplus labor, either underemployed or unemployed workers, who can support the expanded reproduction of electricity capital. Rather than the state investing in its citizens through the provision of a social wage, the state empowers the technical colleges to manage a labor reserve and create a hiring pool for industry. This subsidy to companies relies on a trickle-down logic that increased job creation will benefit the public, taxpayers, and individuals, the latter of whom enter training with the hope of staying in a job at the end.

The effects of the Quick Start program reach far beyond Georgia. The state's aggressive promotion of its workforce drives a low-wage environment in competition with other parts of the US. Without addressing climate change as the reason that has spurred electric vehicle investment, Georgia Governor Brian Kemp has strived to make the state "the electric mobility capital of America" and a central node in the "battery belt" stretching from Michigan to Georgia. He attributes Georgia's success in garnering new investment to "just letting the market work."[126] Generous state incentives shape the market, as do anti-union laws discussed in chapter 4. Instead of standing up for workers to ensure broadly shared prosperity in the energy transition, public resources serve electricity capital.

New energy industries are not the only beneficiaries of state training resources. In 2022, I visited a technical college training program for lineworkers who maintain utility transmission and distribution systems. My host, Emily, described the genesis of the program when Georgia Power came to the Technical College Economic Development department and outlined a need for four hundred new lineworkers. The college customized a ten-week, noncredit program that could be tailored to the specific need, timeframe, and location Georgia Power requested to fast-track training. The programs are available at seven technical colleges across the state, and Georgia Power supports two additional Electric Utility Technology programs. Emily explained, "They have their own internal training program, Klondike, to where they train their line workers . . . and so when they came to us, they basically wanted to mimic that." Georgia Power also gave the college its proprietary curriculum, trucks and other equipment, and leant two instructors to get the program off the ground. "They're being taught the Georgia Power way, if you will. . . . [I]t's a brilliant strategy," enthused Emily. The college benefits from donated employer resources that state and federal agencies lack, and this partnership helps to ensure that graduates most likely have a job at the end. At the time of our meeting, eight of the sixteen graduates had already received job offers from Georgia Power and six others from different regional utilities.

The question of why Georgia Power would rather loan their curriculum to the technical college instead of having workers complete their training in-house is something Emily has thought a lot about. She supposed that "instead of having to spend four, five weeks on training someone with no knowledge . . . it could enable [the utility]'s training folks to sort of expedite their training process so that when students come through here, they can start at a more advanced level, as they've already had the entry level training." In other words, rather than taking a risk to hire an inexperienced worker and help train them on the job, Georgia Power delegated this preparatory work to the technical colleges. While the utility offered their curriculum, equipment, and the time of their trainers to students in the class, it also benefits from the state and trainees taking on a financial and temporal commitment that it would otherwise be required to internalize or share with unions. As a noncredit program, participants cannot receive financial aid, and those who are not eligible for WIOA, which included 75 percent of participants in the program cohort I visited, must pay fees that total $7,500. Workers and the state take on the labor of social reproduction, which becomes another aspect of life-time that sustains the electric utility.

The technical college program has better placement rates than past Georgia Power partnerships. Trey described a partnership he had worked on between Georgia Power and the Urban League. When it began, he was excited.

> Because those are excellent jobs, pays well. Retirement is great. Great company, right? So, what was the challenge there? Well, first of all, they're not that second chance friendly [for people with a criminal record]. . . . They are no tolerance for smoking, that's tobacco and of course marijuana. And then there were some . . . educational requirements that folks had to have as well. . . . I will tell you for about two years, we worked through this process with Georgia Power . . . and I probably ended up getting maybe two or three people through, right? I mean, we did information sessions where a hundred folks will show up because we said we have a Fortune 500 company who's looking for a lineman, we have a training program and a hundred folks will show up, but then we will continue to vet them and then that would narrow down because they didn't meet the requirements, continue to vet them and narrow it down even more.[127]

Figure 5.1
The lineworker training field at a Georgia technical college campus. September 2022. Photo by author.

The disparities in outcomes between the technical college and the Urban League programs are significant in the share of people placed into employment. While intended to create equitable on ramps to employment, these prehire training programs have mixed success even for workers with the time to undertake the work of finding work. Further, training does not address company policies that create barriers to employment. Recognizing some of these limitations, in spring 2025, Georgia Power launched a paid, four-week lineworker entry program with BLOC Resources in metro Atlanta that compensates participants for their time while the utility screens people for its full-time lineworker apprenticeship job.[128]

Through these myriad programs, it is important to consider the relationship between public and nonprofit agencies and the utility. The utility can seize time beyond the wage as it externalizes training to individuals and the public. Training thus becomes another circuit of unvalued time that is both embedded in and enables the expanded reproduction of electric life.

Conclusion

Training is the means to make workers. It has long been a strategy to equip workers to enter the workforce and transition to new employment in periods of economic upheaval. With the rise of neoliberal policy that emphasized personal investment in human capital, training for work was increasingly delegated to the public, families, and individual. Workers, who may have to pay tuition to fund training, or taxpayers, who subsidize facilities and curriculum design for employers, bear the risk. Applying feminist urban political ecology to follow the life-times embedded in energy infrastructures requires asking how electricity capital appropriates the social reproductive time, cooperation, ambition, and skills that people develop at and outside the worksite.[129] Training can become part of a disciplinary strategy to force young people, dislocated workers, and formerly incarcerated people, among others, who are un(der)employed or marginalized within the labor market to enter training in exchange for necessary social services. There is a need for education and upskilling to equip workers for necessary jobs to build and staff the energy transition and paid models codesigned with workers and unions such as registered apprenticeship can support working people. Rather than using training to compel work and make labor cheaper, models for broad, occupational on-the-job training through apprenticeship or public employment ensure that workers have a voice on the job, contribute to job quality and retention, and invest in individual and collective means of reproduction to strengthen the social wage.

Conclusion: Stealing Back Time to Build the Future

On March 14, 2019, former Vice President Al Gore and Reverend William Barber, cochair of the Poor People's Campaign, convened the "Moral Call for Action on Climate Change" at the Ebenezer Baptist Church in Atlanta. The evening started a three-day workshop that brought together participants from across the world for a training with the Climate Reality Project on effective communication about climate action. Climate activists queued in the waning sun of the spring evening outside the church and slowly filed into the modern sanctuary—built to serve a growing congregation from Dr. Martin Luther King Jr.'s historic church across the street. As the evening proceeded, a circle of faith leaders gathered on stage to offer prayers and guidance at the intersection of spiritual and environmental organizing. Their words—some soft and insistent, others thundering with righteous indignation—reverberated through the loudspeakers. A jumbotron hanging above the altar magnified their images wreathed in the yellow light of electric chandeliers. Reverend Gerald Durley began with an invocation as an Atlanta resident, civil rights leader, and board member of the faith-based climate advocacy coalition, Interfaith Power and Light, preaching on the arc of his lifework from civil rights to climate justice. He counseled:

> One of the things we learned in the Civil Rights Movement [with] Dr. King in '59 and 1960 whenever we would come together like this, [is] that there were two times in the Greek, two ways to tell time. One was called Chronos. Chronos means that there are sixty seconds in a minute, so many minutes in an hour, and when you knew that, you'll say chronologically, "we'll do this." But tonight is not a Chronos time, tonight is Kairos. Kairos means there is no time, there are no minutes, there are no hours, but we are divinely inspired to be here holding on to one and other and that we will go forth from this place and save Mother Earth.

The abstract time of Chronos—as Reverend Durley defines it—governs the routines of everyday life, rationalizes slow movement on climate change, obscures the hierarchies that define lived inequalities, and builds a chronological and progressive order for proper action within hegemonic social relations. This sense of chronological time that came to govern labor- and life-time with the proliferation of clocks, energy, and technologies that now define electric life impose a linear sense of progress, locating an imagined future that is somehow more "sustainable" and more just on the horizon as a goal that is "not yet" possible but will come to pass through slow and thus, still profitable, changes to the energy system. The moral imperative for concrete and immediate action on climate change and social injustice ruptures the temporality of Chronos.

Reverend Durley's invocation was followed by remarks from a rabbi and imam, who interpreted the need for climate action within their scripture, community activists, and former Vice President Gore, who spoke as a Southern Baptist to challenge the weaponization of faith in Southern politics that justifies climate change denial. Reverend Barber offered a closing sermon for the evening in which he denounced elected leaders who fail to act in the face of environmental and social injustice. He preached, "Jesus even said in his opening sermon, his evangelical sermon, the sermon that defines what evangelicalism is, which is not are you for prayers in school? are you against women? are you against taxes? are you for guns?" Rather, when you claim to be evangelical it means "to preach good news to the *poor* . . . those who have *been made* poor, who are *hurt* because they are poor, are *poisoned* because they are poor, and who are *stepped on* because they are poor." Moral deeds are those in solidarity with the poor. He repeated as a refrain "the right time to do good, is right now." Now is the right time to end poverty and environmental racism and to address climate change. Doing good work in the South, is even more important, he continued, because "not only has the South been the place of some of the worst hatred it has also been the birthing ground of some of the best revolutions."[1]

Reverend Barber's encouragement to learn from the history of Black struggle against slavery, Jim Crow, and environmental racism pushed the audience to consider the organization of institutions that have perpetuated environmental injustice and sustained the logics of commodification and devaluation. His insistence for everyday action in the tradition of the Black freedom movement evokes the work of Ella Baker who argued forcefully for a feminist praxis of organizing.[2] Less a moment of radical rupture, Baker described that to fight against "capitalism and imperialism" is a process of building relations with neighbors, friends, coworkers, classmates, fellow union members, and churchgoers until "we can get people to recognize that they themselves have to make the struggle and have to make the fight for freedom every day in the year, every year until they win it."[3] The everydayness of popular democratic politics effaces the hegemonic operation of Chronos, wearing away the ordinariness of racial capitalism that steals time from poor and racialized people until "people are coming to realize that, in the final analysis, they are the source of the nation's real worth and power."[4] To enact a relational politics of energy democracy and build coalitions across difference is not only about understanding and protesting injustice but working together to build political alternatives to the utility and Commission that sustains it.

Electricity is a vital need for a healthy and dignified life, necessary to care for ourselves and our families, friends, and neighbors. Georgia Power helped to make electricity integral to everyday life to sell more power. Today, to address climate change and racialized poverty, we must reevaluate how power is made and used and the work involved at every step of the process. This study of Georgia Power concludes with a provocation that asks: How can the utility use its public trust to do good *right now*? To answer this question, I briefly reflect on the methodological lessons of this book. Second, I turn to the key demands raised by organizations in Atlanta throughout the utility's history and put these into conversation with broader institutional critiques of regulated utilities. These policy changes offer a platform for reform that could begin to repair the harms inflicted by corporate control of electrification. The current system of power provision in Atlanta (and much of the US) that privatizes electricity is failing to address the intertwined crises of climate change and affordability, and public regulatory institutions have not stepped up to regulate aggressively in the public interest. Yet, as this book documents, regulation still leaves much to be desired. Public institutions have sustained private electric providers since the nineteenth century and across the Southern economy more broadly, state capacities were organized to advance "industrial prosperity" and maintain access to cheap labor.[5] Rather than reforming the private utility system, we must also ask how a truly public power system could serve the public interest. I close in making the case for energy democracy that embraces public power as a prefigurative politics of building the social wage.

Finding the Times of Electric Life Through Feminist Urban Political Ecology

Electric Life has considered how the investor-owned electric utility in Atlanta fuels the networks of power and difference that shape the built environment, labor, training, and household reproduction to extract value from life-times interconnected through the electric grid. Value, expressed as money, is at the core of social relations within capitalism. The interactions we have with others are mediated through relations based on money, which is necessary to pay for food, clothes, rent, a phone, and the light bill. These commodified demands require people to keep working and, in the process, reproduce the very relations of racism, sexism, and ableism that devalue necessary care work, which may or may not be seen as worthy of a wage. The value of the utility's business comes from regulation through the state Public Service Commission that secures the utility's monopoly control over its service territory and guarantees both a customer base for and rate of return on its product. This political organization of electricity protects the utility monopoly and, in so doing, allows it to harness the unrecognized work of social reproduction in electric life. Electricity rewires everyday life and makes possible the mundane routines of household reproduction, commuting, work, and training. Providing for these quotidian demands for power is the basis of the utility's public charter. Yet, the Commission regulates not with the goal of providing universal electric service to the public, but rather to sustain the private company that sells electricity. As a result, corporate and state power collaborate to maintain the privatized control of a vital public good. The utility operates within and uses its authority to advance a model of economic development that drives racialized, uneven development in Atlanta.

Feminist urban political ecology is a lens to study how the energy system naturalizes, invisibilizes, and appropriates labor- and life-time. The utility seizes surplus labor in paid work and draws into its reach nonstandard moments of social reproduction. Across the chapters, I documented how as a member of the urban regime, Georgia Power encouraged household and industrial energy consumption and coordinated the expanded reproduction of Atlanta to grow electricity demand, promoting a regulatory environment that benefitted capital through low-cost energy, low wages, and low taxes that pushed the cost of social reproduction onto individuals and families. With the consent of the Commission, the utility raises rates, accepting customers suffering as they manage without heat or air conditioning or make sacrifices in their household budgets to pay the light bill. The monopoly granted to Georgia Power by the Commission over a captive service territory enables it to collateralize demand for electricity as a financial asset backed up by the daily needs for power in electric life to borrow for capital projects and continued growth. Supplying power to Atlanta to fuel this growth always depended on water, coal, and other energy resources drawn from far outside the city's limits, which the utility exploited despite the costs to human and environmental health. Its revenue increased in willfully neglecting, violating, and lobbying to neuter environmental law even as coal pollution contributed to asthma, respiratory illness, and cancer, pushing sick individuals and their families to take on these costs. At the same time, the utility's continued use of fossil fuels externalizes the costs associated with a changing climate. On and off the job, the utility exploits the segmentation of the labor market originally founded in a racial and gender hierarchy that reverberates through the continued division of jobs seen as skilled and unskilled. Unions are pushed into unequal alliance with the utility as they fear the loss of their jobs. Their organizing to protect union members from the absence of a social wage juxtaposes with the experience of trainees and prospective Georgia Power workers who take on the "the work of finding work" through self-improvement in training. Feminist urban political ecology zooms into the times of social reproduction to see these interdependent and uneven social processes for how they produce the utility and the wider processes of urbanization in which it is embedded. In making visible the time of social reproduction that sustains the utility, I document the unconsidered and unvalued life-times in the regulatory and legislative decisions that govern public resources like ratepayer funds.

There is a need to change the energy mix, and the economic system it powers defined by racialized uneven development in Atlanta that allows shareholders to benefit from the high rate of return offered by the Commission while the majority face stagnating pay, increasing bills, and an unfunded social welfare state. This becomes all the more important as the future is made less certain and less comfortable by corporate enclosure of the environment to continue fueling climate change.[6] Achieving justice within the existing energy system necessitates democracy—building power to ensure collective determination over resources and time as "*the* resource of life."[7]

Reforming the Utility for the Future

For over a century, Atlantans have questioned the privatized control that Georgia Power and its predecessors have over the grid. This critique has come from a range of perspectives including the antimonopoly, white supremacist populists of the early twentieth century; Black customers who testified at the Commission about unequal treatment on the Jim Crow streetcars; employees within the City of Atlanta moving forward to implement its 100 percent clean energy plan; environmental justice advocates seeking relief from poisoned air and water; union members striking for better wages; and young and unemployed people training for weatherization jobs. Across this history, race, class, and geography have divided social movements as white people refused to see their exploitation by the utility as related to racism and the devaluation of labor and life within racial capitalism. And yet, powerful solidarity movements—most recently mobilized around energy justice and energy democracy—have brokered conversations that cross gender, race, class, and age, the suburbs and the city, clean energy industries, and the low-income consumer advocates and bring together neighbors, coworkers, union members, classmates, and people of faith to ask what people want to see from their power company. These conversations begin with listening and discussions between Georgia Power customers about their intimate needs of electric life. In understanding how people's time is spent to afford their light bills, conserve energy, and organize their neighbors around Commission proceedings, people begin to think together about how this surplus—their time—could be otherwise invested to build infrastructure that supports collective needs.

Contemporary energy democracy organizing draws from a long history of environmental, labor, and consumer movements and the Black freedom movement to demand racial and economic justice in the transition to clean energy. Advocates continue to raise many of the same demands issued by the Georgia Power Project more than half a century ago.[8] These enduring critiques illustrate long-standing failures of the utility, and, in the face of climate change, take on new urgency. Many of these policies are already in place in other states. The utility could implement these changes for which customers and advocates, ranging from the AARP to the Sierra Club and the City of Atlanta, have long advocated on their own. The Commission could also take a more active role to consider disparities in affordability, disconnections, environmental justice, climate action, and the future of the energy workforce.[9]

These wide ranging, yet specific procedural and programmatic changes to reform the utility and Commission to respond to the needs of Georgians now and in the future are worth repeating. These unmet demands raised time and again in IRPs and rate cases include *clean energy* and a fully decarbonized grid by 2035 in line with Atlanta's 100 percent clean energy plan. This would require stopping buildout of new gas infrastructure, phasing-out existing fossil resources, and investing in renewable energies, including through distributed solar, storage, and demand-response. Energy justice advocates and Atlanta residents have lobbied the city and Commission to argue distributing energy resources must be intentional and provide *reparations* to remedy harms of segregation, redlining, gentrification, and environmental racism that inequitable energy infrastructure deepened. Georgia Power has already begun to evaluate how to incorporate distributed energy into its rate base. Recognizing this as a resource that serves the grid, distributed generation procurement could be accelerated and targeted to low- and fixed-income households through community solar subscription programs that lower the cost of energy or residential solar and storage that the utility owns, operates, and manages through virtual power plants. At the same time, the utility could expand energy efficiency, including targeted residential programs to improve household comfort and reduce energy burden, peak demands, and the need to build additional generation.

Steps to finance clean energy must also include programs to guarantee *universal affordability* and *end shut-offs for nonpayment*. The utility could expand limited, income-qualified discounts into percent-of-income payment plans that index utility bills to household income.

These measures gain additional force if regulation is restructured to incentivize a just and equitable energy transition. The rate of return could be indexed to performance in line with decarbonization and equity targets, rather than capitalization, to incent the utility to act for the greatest good for the greatest number. This would deprioritize business recruitment as a trickle-down approach to economic growth and reward investments, policies, and programs that yield broadly shared prosperity.

Each of these changes threatens to disrupt the employment patterns embedded in the existing grid. Coal-fired power plants, coal ash ponds, and other legacy infrastructure must be remediated to ensure *environmental justice*, while protecting communities and workers adjacent to these waste sites. The Commission is outspoken in their support of workers in Commission proceedings, and this should be reflected in policies to ensure a *just transition*. Utility, local, state, and federal incentives have fueled clean energy manufacturing in Georgia, and these investments should serve a broader public interest, including family-supporting wages and benefits, protecting workers' rights to organize and their right to collectively bargain, and prioritizing worker, community, and environmental health. In turn, projects can support paid on-the-job training to begin to address racialized, classed, and gendered segmentation in the energy industry. These common demands for equity at work are organized around the growth of clean energy, which must be evaluated in the context of its effects across global supply chains. These reforms also point to a broader critique of the Southern development agenda that Georgia Power helped to sustain, anticipating how it must shift to build democratic institutions that support working people through the provision of a social wage.

These shifts could gain additional force through *electoral and procedural reform to ensure representation*. In Georgia this requires changing the system of voting for commissioners to correct the systematic underrepresentation of Black voters. Campaign contributions to Commission candidates from utility employees, lawyers, and their family members should also be disallowed. Georgia could also reinstate the Office of Consumers' Utility Counsel.

Achieving a just energy system touches on a much broader set of questions tied to reparations for racial discrimination and climate and environmental injustice, undoing institutions and laws that undercut the power of working people to join together and collectively bargain, and public investment in affordable housing, healthcare, transit, and energy as part of a social wage. Recognizing these interdependencies between healthcare, housing, and energy policy does not mean utilities are off the hook. Rather, energy democracy interrogates the function of the utility and the broader state apparatus that supports in asking who electricity is for; how it is priced, billed, and produced; who owns it; and why it is organized this way to "plant the seeds for a larger movement for structural change."[10]

Calls for reform of the investor-owned utility are rooted in the premise that this quasi-public institution that connects 70 percent of households in Georgia is a necessary infrastructure to build a clean energy future. Shifting the utility to live up to a public mandate acknowledges that the public already sustains the private power company. Regulation makes electricity profitable. Ordinary people are part of and pay for the electric grid. Yet, the long history of unmet demands from Georgia Power reflects that policy and programmatic reform to achieve clean and affordable electricity and universal access to power may not be enough to disrupt the quest for value by a profit-oriented corporation. Democratic governance of the utility will require institutional reform.

Decommodifying Electricity to Build a Social Wage

The investor-owned utility is a paradigmatic instance of the use of state power to serve private interests. This pattern would become ascendent with the rollback of already-limited New Deal regulation of industry in the shift to a neoliberal order. In the South, the white supremacist model of racial capitalist development based on an emaciated vision of the public interest has longer roots. Yet, over decades, public institutions have been created and transformed through iterative organizing of mass social movements to fight industry capture, protect the environment, recognize and honor global interdependencies, refashion collective infrastructures, build public power, and invest in community care.[11] Throughout the history of electric utilities, changes have come only through collective action to transform these collective institutions established in and through the state. Democratizing energy must entail an everyday politics of social reproduction that begins with asking what we need energy for and guarantees affordable and clean energy as part of a social wage. This starting point can recast state power not to serve industrial growth but rather invest ratepayer funds and tax dollars in services for the public good.

Energy democracy is hard to imagine in the second Trump administration as state institutions are actively being dismantled. At the same time, facing what Georgia Power warns is the largest growth in power demand since the advent of air conditioning with data centers, members of the public and Commission staff have questioned whether the public should be responsible for building massive amounts of new fossil generation to meet this projected demand.[12] Others have questioned Georgia Power's high bills relative to their neighbors served by electric co-ops. These commenters fundamentally question the growth mandate and profit motive of the utility.

Many have looked to the New Deal era when the Rural Electrification Administration and the TVA quickly electrified rural regions of the US, as a model alternative that demonstrates how a publicly controlled energy system could rapidly scale to meet the urgency of climate change.[13] The TVA, electric co-ops, municipal utilities, and a federal power agency all operate in Georgia.[14] These public and nonprofit models have been shaped through their intertwined coexistence with investor-owned utilities. In many ways they are not perfect—representing a deficit of public participation and oversight, a lack of diversity on governing boards, and continued investment in fossil fuels. Nonetheless, public power institutions that serve 30 percent of Georgia are a functional example of how the profit motive can be removed from utility service and public governance implemented.

Public power campaigns launched in cities and states across the US have investigated the public takeover of private utilities or renegotiated the terms under which investor-owned utilities operate.[15] These efforts have secured new renewable energy and energy efficiency investments, programs to address energy burden, and other public priorities. A broad coalition of unions and environmental advocates won the Build Public Renewables Act in New York to allow the state-owned New York Public Power Authority to build and own clean energy projects.[16] After devastating wildfires caused by negligence and poor maintenance of Pacific Gas and Electric infrastructure in California, elected leaders demanded overhaul of governance and operation of utility at the threat of a takeover by the state.[17] California is also considering the public takeover of refineries to manage production of gasoline as refineries close in response to climate policy.[18]

These instances of muscular state intervention reimagine how state power that makes the utility can also be used to regulate, own, or operate electricity infrastructure to serve poor and unemployed people, renters, single parents, and elders who are currently marginalized within a system that commodifies electric life. In his historical review of public power, Sandeep Vaheesan offers one model for how public power could be achieved through congressional action to establish new federal power agencies under the direction of community and worker boards committed to decarbonization of the grid.[19] His framework reimagines the role of the state as a financing entity and governing body as well as people's relationship to the utility by drawing on the expertise of workers and community members living with energy infrastructure to share in the governance of a community owned and operated energy asset.

Implementing any such proposals to expand public power would have to wrestle with the contradictions of state power forged through Atlanta's urban regime and regional bloc in Georgia that, despite efforts of progressive social movements in the city, has been and remains vested in white supremacy, anti-LGBTQ repression, policing of migrants and people of color, and militarism and divestment from a social wage, including affordable electricity and a healthy environment. Violent institutions organized through the state reinforce racial capitalism as a hierarchical ordering of people. Yet, the state is not a monolith; instead, it is a collection of people, conventions, laws, and infrastructures. State regulatory institutions and the organization of state capacities arise from contestation through a shared project of governance. Since Reconstruction, civic leaders who would form the urban regime in Atlanta have mobilized the fiscal capacities and ideological authority of the state to promote a narrow vision of development—contested through the long civil rights movement as Black people excluded from the formal mechanisms of government demanded representation. Recognizing, as Reverend Barber did, the contradictions of Southern political histories, we must hold these pieces in tension to devise energy and climate strategies for a safe future based on the improvement of collective infrastructures and new relations of public power that cultivate shared respect and investment of time. It requires radically altering the balance of power organized through the state government and Atlanta's urban regime and using "unrestrained imagination" to envision how the electric grid as a collective project could support community and household reproduction to reconstitute the eviscerated social wage.[20] To achieve this requires democratic action by workers, environmental justice communities, poor people, and others marginalized within the networks of the existing electric grid to change the mechanisms of state power and the operating definition of the public interest and to devise systems that provide for their daily needs.

Organizing begins with relationships as people find each other myriad settings from hearings inside the Public Service Commission to rallies outside; meetings on Zoom and in church basements; from happy hours to the AFL-CIO Labor Day cookout. Solidarity against injustice and collaborations to build a more just future are forged between unions and faith groups, environmental justice and reproductive justice organizations, and energy equity organizers and environmentalists.[21] Coalition work requires historical reevaluation to recognize and redress past harms and define energy democracy collectively to build and nourish connections and dismantle the hierarchies that differentially and discriminately value time. If this book has in part been a meditation on the relational processes of valuation and devaluation that shape Atlanta, it has also shown the connections that reach between "differentially laboring people" and the intimacies that follow through electric infrastructure in connecting people and places together.[22] Energy democracy is a starting point to reorganize the grid in ways that envision nonfatal and life-affirming relations that reject the devaluation of time.[23] Disrupting the quotidian rhythms of Chronos scaffolded by electricity challenges the temporality of the utility as an infrastructure of power. Counterhegemonic temporalities move beyond the reproduction of value to realize the uses of electricity to build healthy and equitable lifeways in Atlanta. As Bobby Seale argues, collective action begins with seizing the time to "do something about the wretched conditions that exist, the poverty that exists, and the unjust wars that are waged in the world. The whole future must be directly related to what we, the people, want. . . . We must all start seeing ourselves beginning to seize the time."[24]

Notes

Introduction

1. City of Atlanta Mayor's Office of Resilience, "Clean Energy Atlanta: A Vision for a 100% Clean Energy Future," 2018, http://100atl.com/; Wei Lu and Alexandre Tanzi, "In America's Most Unequal City, Top Households Rake In $663,000," *Bloomberg*, November 21, 2019, https://www.bloomberg.com/news/articles/2019-11-21/in-america-s-most-unequal-city-top-households-rake-in-663-000; Ariel Drehobl, Lauren Ross, and Roxana Ayala, "How High Are Household Energy Burdens? An Assessment of National and Metropolitan Energy Burden across the United States" (American Council for an Energy-Efficient Economy, September 10, 2020).

2. Partnership for Southern Equity et al., "Rally to #FightTheHike," Facebook, September 30, 2019, https://www.facebook.com/events/380239829321214/?active_tab=discussion.

3. Denise Fairchild and Al Weinrub, eds., *Energy Democracy: Advancing Equity in Clean Energy Solutions* (Washington, DC: Island Press, 2017).

4. David Roberts, "This New Organization Wants to Remake PUCs for the Energy Transition," Volts, accessed December 14, 2024, https://www.volts.wtf/p/this-new-organization-wants-to-remake.

5. Gretchen Bakke, *The Grid: The Fraying Wires Between Americans and Our Energy Future* (New York, NY: Bloomsbury, 2016), 78.

6. Richard F. Hirsh, *Power Loss: The Origins of Deregulation and Restructuring in the American Electric Utility System* (Cambridge: MIT Press, 1999), 31.

7. Werner Troesken, "Regime Change and Corruption: A History of Public Utility Regulation," in *Corruption and Reform: Lessons from America's Economic History*, ed. Edward L. Glaeser and Claudia Goldin (University of Chicago Press, 2006), 259–281.

8. Gabriel Chan and Alexandra B Klass, "Regulating for Energy Justice," *New York University Law Review* 97 (2022): 1426–1506.

9. US Energy Information Administration, "US Energy-Related Carbon Dioxide Emissions, 2023," April 25, 2024, https://www.eia.gov/environment/emissions/carbon/; U.S. Energy Information Administration, "US Energy Facts Explained," July 15, 2024, https://www.eia.gov/energyexplained/us-energy-facts/.

10. David Roberts, "The Key to Tackling Climate Change: Electrify Everything," Vox, Ocotber 2017, https://www.vox.com/2016/9/19/12938086/electrify-everything.

11. *PSC Dockets #56002 & #56003 Georgia Power Company 2025 Plans & Applications*, 2025, Georgia Public Service Commission, https://www.youtube.com/watch?v=Yyf7f8IsHCE; Sonal Patel. Georgia Power to Keep Coal, Gas Power Plants Running Longer as Demand Climbs. Power Magazine, February 5, 2025, https://www.powermag.com/georgia-power-to-keep-coal-gas-power-plants-running-longer-as-demand-climbs-2/.

12. Brett Christophers, *The Price Is Wrong: Why Capitalism Won't Save the Planet* (London New York: Verso, 2024).

13. US Energy Information Administration, "US Energy Facts Explained."

14. On definitions of production see Mariana Mazzucato, *The Value of Everything: Making and Taking in the Global Economy* (New York: PublicAffairs, 2018).

15. Cindi Katz, "Vagabond Capitalism and the Necessity of Social Reproduction," *Antipode* 33, no. 4 (2001): 709–728; Beverley Mullings, "Caliban, Social Reproduction and Our Future Yet to Come," *Geoforum* 118 (2021): 150–158.

16. From this perspective, time at home (time of life, social reproduction) and work (labor-time, production) are not two separate spheres but interrelated; labor is nested within and relies on the often unrecognized "life's work" of social reproduction. Feminist analyses of "life's work" and postcolonial feminist studies of "life-time" and "life work" attempt to show the relations within and beyond the reach of capital that nonetheless sustain labor. For efforts to move beyond the binary of production/reproduction see: Angela Y. Davis, *Women, Race & Class* (New York: First Vintage Books, 1983); Max Andrucki et al., "Beyond Binaries and Boundaries in 'Social Reproduction,'" 2017, https://www.societyandspace.org/forums/beyond-binaries-and-boundaries-in-social-reproduction. On life's work, see Katharyne Mitchell, Sallie A. Marston, and Cindi Katz, "Life's Work: An Introduction, Review and Critique," *Antipode* 35, no. 3 (2003): 415–442; on life-work, Mullings, "Caliban, Social Reproduction and Our Future Yet to Come"; on life-time, Neferti X. M. Tadiar, *Remaindered Life* (Durham: Duke University Press Books, 2022).

17. Nikki Luke and Matthew T. Huber, "Introduction: Uneven Geographies of Electricity Capital," *Environment and Planning E: Nature and Space* 5, no. 4 (2022): 1699–1715.

18. Willie Jamaal Wright, "The Morphology of Marronage," *Annals of the American Association of Geographers* 110, no. 4 (2020): 1134–1149; Nikki Luke, "Powering

Racial Capitalism: Electricity, Rate-Making, and the Uneven Energy Geographies of Atlanta," *Environment and Planning E: Nature and Space* 5, no. 4 (2022): 1765–1787; Rachel Goffe and Nikki Luke, "What Does Capital Consume? Racial Capitalism and the Social Reproduction of Surplus People," *Environment and Planning A: Economy and Space* 56, no. 4 (2024): 1311–1319.

19. Cedric Robinson, *Black Marxism: The Making of the Black Radical Tradition* (Chapel Hill, NC: University of North Carolina Press, 2000); also Lisa Lowe, *The Intimacies of Four Continents* (Durham, NC: Duke University Press, 2015), 149–150; Olúfẹ́mi O. Táíwò, *Reconsidering Reparations* (New York: Oxford University Press, 2022).

20. Interview with environmental advocate, April 17, 2024; Beginning in the 1990s, about half of states deregulated electricity. Conor Harrison and Shelley Welton, "The States That Opted Out: Politics, Power, and Exceptionalism in the Quest for Electricity Deregulation in the United States South," *Energy Research & Social Science* 79 (2021): 8.

21. Chandra Childers, "Southern Economic Policies Undermine Job Quality for Auto Workers: Rooted in Racism and Economic Exploitation" (Economic Policy Institute, September 4, 2024), 1, https://www.epi.org/publication/rooted-racism-auto-workers/; also, Regina S. Baker, "The Historical Racial Regime and Racial Inequality in Poverty in the American South," *American Journal of Sociology* 127, no. 6 (2022): 1721–1781.

22. Conor Harrison and Jeff Popke, "'Because You Got to Have Heat'—the Networked Assemblage of Energy Poverty in Eastern North Carolina," *Annals of the Association of American Geographers* 101, no. 4 (2011): 949–961.

23. Nelson Lichtenstein, *State of the Union: A Century of American Labor* (Princeton, N.J.: Princeton University Press, 2003), 11; Lichtenstein's explanation of the social wage is defined in reference to Gøsta Esping-Andersen, *The Three Worlds of Welfare Capitalism* (Princeton, NJ: Princeton University Press, 1990).

24. Last updated December 2024. E2, "REPORT: Recent Projects in Georgia Push Clean Energy Jobs over 82,000 As Sector Adds Jobs 4X Faster Than GA Economy," December 11, 2024, https://e2.org/releases/report-recent-projects-in-georgia-push-clean-energy-jobs-over-82000-as-sector-adds-jobs-4x-faster-than-ga-economy/. Georgia does not have a renewable portfolio standard or net-metering to encourage solar adoption.

25. This applies to all municipal operations, including the city-owned Hartsfield-Jackson Airport and all electricity used within city limits.

26. Rhodium Group and MIT Center for Energy and Environmental Policy Research, "Clean Investment Monitor," 2025, https://www.cleaninvestmentmonitor.org/database.

27. Southern Company, "2024 Annual Report," April 1, 2025, 56, https://s27.q4cdn.com/273397814/files/doc_financials/2023/ar/2024-annual-report.pdf.

28. Georgia Power, "Facts & Figures," 2024, https://www.georgiapower.com/about/company/facts-and-figures.html; Southern Company, "Our Companies," 2024, https://www.southerncompany.com/about/our-companies.html; Conor Harrison, "Electricity Capital and Accumulation Strategies in the U.S. Electricity System," *Environment and Planning E: Nature and Space*, 2020.

29. James C. Cobb, *The Selling of the South: The Southern Crusade for Industrial Development, 1936–1990* (Urbana, IL: University of Illinois Press, 1993); Bruce J. Schulman, *From Cotton Belt to Sunbelt: Federal Policy, Economic Development, and the Transformation of the South 1938–1980* (Durham, NC: Duke University Press, 1994); Gavin Wright, *Old South, New South: Revolutions in the Southern Economy Since the Civil War* (Baton Rouge, LA: Louisiana State University Press, 1997).

30. H. C. Nixon, "The Changing Political Philosophy of the South," *The Annals of the American Academy of Political and Social Science* 153 (1931): 248; also Paul Gaston, *New South Creed: A Study in Southern Mythmaking* (Montgomery, AL: NewSouth, Incorporated, 2001).

31. Clyde Woods, *Development Arrested: The Blues and Plantation Power in the Mississippi Delta* (London: Verso, 1998), 26; Clyde Woods, *Development Drowned and Reborn: The Blues and Bourbon Restorations in Post-Katrina New Orleans*, ed. Laura Pulido and Jordan Camp (Athens, GA: University of Georgia Press, 2017); on race and hegemony also, Bobby M. Wilson, *America's Johannesburg: Industrialization and Racial Transformation in Birmingham* (Athens, GA: University of Georgia Press, 2019).

32. Clarence N. Stone, *Regime Politics: Governing Atlanta, 1946–1988* (Lawrence, KS: University Press of Kansas, 1989).

33. Maurice J. Hobson, *The Legend of the Black Mecca: Politics and Class in the Making of Modern Atlanta* (Chapel Hill, NC: The University of North Carolina Press, 2017), 1; Akira Drake Rodriguez, *Diverging Space for Deviants: The Politics of Atlanta's Public Housing* (Athens, GA: University of Georgia Press, 2021).

34. Kevin Kruse, *White Flight: Atlanta and the Making of Modern Conservatism* (Princeton, NJ: Princeton University Press, 2007), 243.

35. Manning Marable, "Power to the People? Energy and Economic Underdevelopment of Black People in the 'New South,'" *Black Books Bulletin* 7, no. 3 (1981): 10.

36. Canay Özden-Schilling, *The Current Economy: Electricity Markets and Techno-Economics* (Stanford, CA: Stanford University Press, 2021), 15–16; Thomas P. Hughes, *Networks of Power: Electrification in Western Society, 1880–1930* (Baltimore, MD: John Hopkins University Press, 1993); David E. Nye, *Electrifying America: Social Meanings of a New Technology, 1880–1940* (Cambridge, MA: MIT Press, 1992).

37. Sharon Beder, *Power Play: The Fight to Control the World's Electricity* (New York: The New Press, 2003), 24; Richard Munson, *The Power Makers: The Inside Story of America's*

Biggest Business . . . and Its Struggle to Control Tomorrow's Electricity (Emmaus, PA: Rodale Press, 1985).

38. Bakke, *The Grid,* 57. Electric utilities can be considered one of several "select technological enterprises that obtained special treatment in American society." See: Hirsh, *Power Loss*, 31.

39. Valery Yakubovich, Mark Granovetter, and Patrick McGuire, "Electric Charges: The Social Construction of Rate Systems," *Theory and Society* 34, no. 5–6 (2005): 579–612.

40. Bakke, *The Grid,* 67. Characterizing eleven different customer classes with a combined load of 26,640 kw in 1914, Insull noted that by matching industries with different peak demands, the maximum peak reached on the grid was only 9,770 kW. Hughes, *Networks of Power*, 204–217.

41. Özden-Schilling, *The Current Economy,* 17. Outside the US, electricity was often provided by national utilities although efforts to marketize and privatize electricity as part of neoliberal reforms beginning in the 1980s changed the landscape of public ownership. See: Hughes, *Networks of Power*; Christophers, *The Price Is Wrong*; Elizabeth Chatterjee, "The Asian Anthropocene: Electricity and Fossil Developmentalism," *The Journal of Asian Studies* 79, no. 1 (2020): 3–24.

42. Troesken, "Regime Change and Corruption: A History of Public Utility Regulation."

43. Ruth Wilson Gilmore, *Golden Gulag: Prisons, Surplus, Crisis, and Opposition in Globalizing California* (Berkeley, CA: University of California Press, 2007), 78.

44. Hirsh, *Power Loss.*

45. Conor Harrison, "'Accomplished by Methods Which Are Indefensible': Electric Utilities, Finance, and the Natural Barriers to Accumulation," *Geoforum* 49 (2013), 178.

46. James C. Bonbright, *Principles of Public Utility Rates* (New York: Columbia University Press, 1961), 389.

47. Steve Kihm et al., "You Get What You Pay For: Moving Toward Value in Utility Compensation, Part 1, Revenue and Profit" (San Francisco: America's Power Plan, 2015), 4, https://americaspowerplan.com/wp-content/uploads/2016/07/CostValue-Part1-Revenue.pdf.

48. Mark A. Jamison, "Rate of Return: Regulation" (Gainesville, FL: Public Utility Research Center, 2005), 3; Bonbright, *Principles of Public Utility Rates.*

49. Hirsh, *Power Loss*, 46.

50. Beder, *Power Play,* 63.

51. Quote from "Testimony of Preston S. Arkwright Chairman of the Public Policy Committee of the National Electric Light Association before the Federal Trade

Commission" (Washington, DC: National Electric Light Association, January 14, 1930), 66–67, Atlanta History Center, MSS 356, Box 3, Folder 3; on Georgia Power's advertising practices, see Hamilton Frazier Moore, "An Evaluation of the Art and Techniques of Corporate Public Relations as Practices by the Georgia Power Company" (PhD, University of Illinois at Urbana-Champaign, 1965), 21–22; on utility advertising, Richard Rudolph and Scott Ridley, *Power Struggle: The Hundred-Year War over Electricity* (New York: Harpercollins, 1986).

52. Bakke, *The Grid*, 59–60.

53. Brent Cebul, "Creative Competition: Georgia Power, the Tennessee Valley Authority, and the Creation of a Rural Consumer Economy, 1934–1955," *Journal of American History* 105, no. 1 (2018): 48.

54. Taft and Heys, *Big Bets*, 117.

55. Wendell L. Willkie, "Spending for Load Building," *Electrical World* 104 (December 8, 1934): 21.

56. Matthew T Huber, *Lifeblood: Oil, Freedom, and the Forces of Capital* (Minneapolis: University of Minnesota Press, 2013), 23.

57. Cebul, "Creative Competition," 48; Nye, *Electrifying America: Social Meanings of a New Technology, 1880–1940*; Samer Alatout and Chelsea Schelly, "Rural Electrification as a 'Bioterritorial' Technology: Redefining Space, Citizenship, and Power During the New Deal," *Radical History Review* 2010, no. 107 (2010): 127–138; Michelle Mock, "The Electric Home and Farm Authority, 'Model T Appliances,' and the Modernization of the Home Kitchen in the South," *The Journal of Southern History* 80, no. 1 (2014): 75–76; on the New Deal, electrification, and racial disparities in electricity access, see Andrew Needham, *Power Lines: Phoenix and the Making of the Modern Southwest* (Princeton, NJ: Princeton University Press, 2014); Conor Harrison, "Race, Space, and Electric Power: Jim Crow and the 1934 North Carolina Rural Electrification Survey," *Annals of the American Association of Geographers* 106, no. 4 (2016): 909–931.

58. Moore, "An Evaluation of the Art and Techniques of Corporate Public Relations as Practices by the Georgia Power Company," 67.

59. Taft and Heys, *Big Bets*, 131.

60. Casey Cater, *Regenerating Dixie: Electric Energy and the Modern South* (Pittsburgh, PA: University of Pittsburgh Press, 2019).

61. For Green New Deal organizing in Georgia, see "DeKalb Green New Deal," Commissioner Ted Terry, 2024, https://commissionertedterry.com/dekalb-green-new-deal/; on the Green New Deal, Kate Aronoff et al., *A Planet to Win: Why We Need a Green New Deal* (London: Verso, 2019).

62. Huber, *Lifeblood*.

63. Nancy MacLean, "Southern Dominance in Borrowed Language: The Regional Origins of American Neoliberalism," in *New Landscapes of Inequality: Neoliberalism and the Erosion of Democracy in America*, ed. Jane L. Collins, Micaela di Leonardo, and Brett Williams (Santa Fe, NM: School for Advanced Research Press, 2008).

64. W.E.B. Du Bois, *Black Reconstruction in America, 1860–1880* (New York: The Free Press, 1935), 700; Ruth Wilson Gilmore, "Fatal Couplings of Power and Difference: Notes on Racism and Geography," *The Professional Geographer* 54, no. 1 (2002): 15–24; Lichtenstein, *State of the Union*.

65. Nancy MacLean, "Southern Dominance in Borrowed Language," 28; also epilogue in Robin D. G. Kelley, *Hammer and Hoe: Alabama Communists during the Great Depression*, Twenty-Fifth Anniversary Edition (Chapel Hill, NC: The University of North Carolina Press, 2015); Kate Driscoll Derickson, "The Racial Politics of Neoliberal Regulation in Post-Katrina Mississippi," *Annals of the Association of American Geographers* 104, no. 4 (2014): 889–902.

66. Peck and Theodore also demonstrate the racist and authoritarian policies that weaponized the state against marginalized people while promoting economic freedom in defining the historical roots of neoliberalism see Jamie Peck and Nik Theodore, "Still Neoliberalism?," *South Atlantic Quarterly* 118, no. 2 (2019): 245; on the Southern roots of neoliberalism, see Melinda Cooper, "The Alt-Right: Neoliberalism, Libertarianism and the Fascist Temptation," *Theory, Culture & Society* 38, no. 6 (2021): 29–50. Sellers adds "antistatist and deregulatory directions" adopted during President Carter's administration that were seen as a watershed in the shift from the New Deal to neoliberal order were "stubborn, nationally projected norms of Georgia's own political economy and culture." Christopher C. Sellers, *Race and the Greening of Atlanta: Inequality, Democracy, and Environmental Politics in an Ascendant Metropolis* (Athens, GA: University of Georgia Press, 2023), 167; on the shift from New Deal to neoliberal order, see: Gary Gerstle, *The Rise and Fall of the Neoliberal Order: America and the World in the Free Market Era* (Oxford: Oxford University Press, 2022).

67. Melinda Cooper, *Family Values: Between Neoliberalism and the New Social Conservatism* (New York: Zone Books, 2017), 71; Cindi Katz, "Bad Elements: Katrina and the Scoured Landscape of Social Reproduction," *Gender, Place & Culture* 15, no. 1 (2008): 15–29.

68. "Martin Luther King Jr. Speaks with NBC News 11 Months Before Assassination" (Atlanta, GA: NBC News 11, May 8, 1967), https://www.nbcnews.com/video/martin-luther-king-jr-speaks-with-nbc-news-11-months-before-assassination-1202163779741?cid=sm_npd_nn_tw_ma.

69. "PSC's Echols: 'It's Amazing We Ever Finished the [Vogtle] Project at All,'" *The Georgia Virtue*, June 3, 2024, https://www.thegeorgiavirtue.com/georgia-news/pscs-echols-its-amazing-we-ever-finished-the-vogtle-project-at-all/.

70. Personal communication with energy advocate, June 10, 2025.

71. Harrison and Welton, "The States That Opted Out," 8.

72. For an early commentary on these global effects see: Du Bois, *Black Reconstruction in America, 1860–1880*, 634.

73. Barry Bluestone and Bennett Harrison, *The Deindustrialization of America* (New York: Basic Books, 1982), 182.

74. Southern Company, "Imagine 2017 Annual Report," March 20, 2018, 1, https://s2.q4cdn.com/471677839/files/doc_financials/annual2017/2017-AR-Full-PDF-FINAL.pdf.

75. Southern Company, 24.

76. See chapter 1.

77. Mazzucato also critiques marginal choice theory that equates value to price as determined by consumer willingness to pay as a reaction to classical political economy based on the labor theory of value in which value was thought to come from people producing goods and services, or workers. Mazzucato, *The Value of Everything*, 6, 58.

78. Matthew T. Huber, "Energizing Historical Materialism: Fossil Fuels, Space and the Capitalist Mode of Production," *Geoforum* 40, no. 1 (2009): 105–115; Timothy Mitchell, *Carbon Democracy: Political Power in the Age of Oil* (New York: Verso, 2013); Brent Ryan Bellamy and Jeff Diamanti, eds., *Materialism and the Critique of Energy* (Chicago: MCM Publishing, 2018).

79. Andreas Malm, *Fossil Capital: The Rise of Steam Power and the Roots of Global Warming* (London: Verso, 2016).

80. Moishe Postone, *Time, Labor, and Social Domination: A Reinterpretation of Marx's Critical Theory* (Cambridge, UK: Cambridge University Press, 1993), 202.

81. Malm, *Fossil Capital*, 290.

82. Karl Marx, *Capital: A Critique of Political Economy, Volume I* (London: Penguin Books, 1990).

83. My understanding of value comes from scholars who read Marx's work as an immanent critique of value, Postone, *Time, Labor, and Social Domination*; Diane Elson, "The Value Theory of Labour," in *Value: The Representation of Labour in Capitalism*, ed. Diane Elson (London: Verso, 2015); also, Melinda Cooper, "Workfare, Familyfare, Godfare: Transforming Contingency into Necessity," *South Atlantic Quarterly* 111, no. 4 (2012): 643–661; Melinda Cooper and Catherine Waldby, *Clinical Labor: Tissue Donors and Research Subjects in the Global Bioeconomy* (Durham, NC: Duke University Press Books, 2014).

84. Postone, *Time, Labor, and Social Domination*, 25.

85. Stuart Hall, "Gramsci's Relevance for the Study of Race and Ethnicity," *Journal of Communication Inquiry* 10, no. 2 (1986): 24.

86. Mullings, "Caliban, Social Reproduction and Our Future Yet to Come," 152.

87. C. L. R. James, *The Black Jacobins: Toussaint L'Ouverture and the San Domingo Revolution*, 2nd edition (New York: Vintage, 1989); Robinson, *Black Marxism*; Silvia Federici, *Caliban and the Witch: Women, the Body, and Primitive Accumulation* (Brooklyn, NY: Autonomedia, 2014); Maria Mies, *Patriarchy and Accumulation on a World Scale: Women in the International Division of Labour* (London: Zed Books, Ltd., 2014); Lowe, *The Intimacies of Four Continents*; Jodi Melamed, "Racial Capitalism," *Critical Ethnic Studies* 1, no. 1 (2015): 76–85; also Jodi A. Byrd et al., "Predatory Value," *Social Text* 36, no. 2 (2018): 1–18.

88. Davis, *Women, Race & Class*.

89. Leopoldina Fortunati, *Arcane of Reproduction: Housework, Prostitution, Labor and Capital*, ed. Jim Fleming, trans. Hilary Creek (Brooklyn, NY: Autonomedia, 1996); Federici, *Caliban and the Witch*.

90. Nancy Fraser, *Cannibal Capitalism: How Our System Is Devouring Democracy, Care, and the Planet and What We Can Do About It* (London: Verso, 2022), xiv.

91. Myles Lennon, "Decolonizing Energy: Black Lives Matter and Technoscientific Expertise amid Solar Transitions," *Energy Research & Social Science* 30 (2017): 25; also, Cara New Daggett, *The Birth of Energy: Fossil Fuels, Thermodynamics, and the Politics of Work* (Durham, NC: Duke University Press, 2019).

92. Moishe Postone, "The Current Crisis and the Anachronism of Value: A Marxian Reading," *Continental Thought and Theory A Journal of Intellectual Freedom* 1, no. 4 (2019): 38–54.

93. Vaclav Smil, *Energy: A Beginner's Guide* (Oxford: Oneworld Publications, 2006), 103–104.

94. Henri Lefebvre, *The Production of Space* (Malden, MA: Blackwell Publishers, 1991), 347.

95. Nye, *Electrifying America*, 132.

96. Harvey, *Seventeen Contradictions and the End of Capitalism*, 238.

97. Marx observes that "the prolongation of the working day beyond the limits of the natural day, into the night . . . only slightly quenches the vampire thirst for the living blood of labour." Marx, *Capital: A Critique of Political Economy, Volume I*, 367.

98. Digital spaces further facilitate "the economization of heretofore noneconomic spheres and practices" as everything from dating, to tracking workouts, education, and social media can become an investment in human capital. Wendy Brown,

Undoing the Demos: Neoliberalism's Stealth Revolution (Cambridge, MA: MIT Press, 2017) 30–31, 34. Others, portend the collapse of labor as digital technologies render "time permanently available to connect." Franco Berardi Bifo, *After the Future*, ed. Gary Genosko and Nicholas Thoburn (New York: AK Press, 2011), 66. In the post-Fordist, knowledge economy, Lazzarato argues capitalist accumulation exploits "immaterial labor" in the form of "knowledge, life, health, leisure, culture etc." Maurizio Lazzarato, "From Capital-Labour to Capital-Life," *Emphemera: Theory & Politics in Organization* 4, no. 3 (2004): 205. Yet, Cooper and Waldby caution that the application of such theories can lose specificity to understand the power relations that exert pressure on precaritized workers. Cooper and Waldby, *Clinical Labor*, 6–9.

99. Jenny Odell, *How to Do Nothing: Resisting the Attention Economy* (Brooklyn, NY: Melville House, 2019), xii; Yanis Varoufakis, "Quantity to Quality," *NLR/Sidecar*, November 20, 2024.

100. For studies that theorize as energy through the lens of social reproduction see: Huber, *Lifeblood*; James Angel, "Irregular Connections: Everyday Energy Politics in Catalonia," *International Journal of Urban and Regional Research* 43, no. 2 (2019): 337–353; Gabriela Valdivia, "'Wagering Life' in the Petro-City: Embodied Ecologies of Oil Flow, Capitalism, and Justice in Esmeraldas, Ecuador," *Annals of the American Association of Geographers* 108, no. 2 (2018): 549–557; Stefan Bouzarovski, "Energy and Labour: Thinking Across the Continuum," *Progress in Human Geography* 46, no. 3 (2022): 753–774.

101. Ruth Schwartz Cowan, *More Work for Mother: The Ironies of Household Technology from the Open Hearth to the Microwave* (New York: Basic Books, 1985), 13.

102. Susan Strasser, *Never Done: A History of American Housework* (New York: Holt Paperbacks, 2000), 4.

103. Mazzucato, *The Value of Everything*, 91–92.

104. Gargi Bhattacharyya, *Rethinking Racial Capitalism: Questions of Reproduction and Survival* (Lanham, MD: Rowman & Littlefield International, 2018), 154.

105. Mitchell, Marston, and Katz, "Life's Work: An Introduction, Review and Critique," 415, 418.

106. Cooper, "Workfare, Familyfare, Godfare," 645.

107. Neferti X. M. Tadiar, "Life-Times of Becoming Human," *Occasion: Interdisciplinary Studies in the Humanities* 3 (2012): 11.

108. Saska Petrova and Neil Simcock, "Gender and Energy: Domestic Inequities Reconsidered," *Social & Cultural Geography*, 2019, 1–19; Charlotte Johnson, "Is Demand Side Response a Woman's Work? Domestic Labour and Electricity Shifting in Low Income Homes in the United Kingdom," *Energy Research & Social Science* 68 (2020): 101558; Shuchen Cong et al., "Unveiling Hidden Energy Poverty Using the Energy Equity Gap," *Nature Communications* 13, no. 1 (2022): 2456; Mark Kear et al.,

"Home Thermal Security, Energy Equity and the Social Production of Heat in Manufactured Housing," *Energy Research & Social Science* 106 (2023): 103318.

109. Melamed, "Racial Capitalism," 78.

110. Gilmore, "Fatal Couplings of Power and Difference," 16; on slow violence, Rob Nixon, *Slow Violence and the Environmentalism of the Poor* (Cambridge, MA: Harvard University Press, 2011); also chapter 3.

111. Ruth Wilson Gilmore, "Abolition Geography and the Problem of Innocence," in *Futures of Black Radicalism*, ed. Gaye Theresa Johnson and Alex Lubin (London: Verso, 2017), 227.

112. Elson, "The Value Theory of Labour," 171–172.

113. Kacper Szulecki, "Conceptualizing Energy Democracy," *Environmental Politics* 27, no. 1 (2018): 21–41; Becker, Angel, and Naumann, "Energy Democracy as the Right to the City."

114. Shalanda Baker, *Revolutionary Power: An Activist's Guide to the Energy Transition* (Washington, DC: Island Press, 2021), 30; Fairchild and Weinrub, *Energy Democracy: Advancing Equity in Clean Energy Solutions*.

115. Kacper Szulecki and Indra Overland, "Energy Democracy as a Process, an Outcome and a Goal: A Conceptual Review," *Energy Research & Social Science* 69 (2020): 101768.

116. Matthew J. Burke and Jennie C. Stephens, "Energy Democracy: Goals and Policy Instruments for Sociotechnical Transitions," *Energy Research & Social Science* 33 (2017): 35–48; James Angel, "Towards an Energy Politics In-Against-and-Beyond the State: Berlin's Struggle for Energy Democracy," *Antipode* 49, no. 3 (2017): 557–576; John Farrell, "Boulder Voters Say (Again): 'We'll Lead Movement to Energy Democracy,'" Institute for Local Self-Reliance, November 21, 2017, https://ilsr.org/boulder-voters-say-again-well-lead-movement-to-energy-democracy/.

117. James McCarthy, "A Socioecological Fix to Capitalist Crisis and Climate Change? The Possibilities and Limits of Renewable Energy," *Environment and Planning A: Economy and Space* 47, no. 12 (2015): 2485–2502; Shelley Welton, "Grasping for Energy Democracy," *Michigan Law Review* 116, no. 4 (2018): 581–644; Myles Lennon, "Energy Transitions in a Time of Intersecting Precarities: From Reductive Environmentalism to Antiracist Praxis," *Energy Research & Social Science* 73 (2021): 101930.

118. Eric O'Shaughnessy et al., "The Impact of Policies and Business Models on Income Equity in Rooftop Solar Adoption," *Nature Energy* 6, no. 1 (2021): 84–91.

119. Nikki Luke and Nik Heynen, "Community Solar as Energy Reparations: Abolishing Petro-Racial Capitalism in New Orleans," *American Quarterly* 72, no. 3 (2020): 603–625; Kimberly C. Yazzie et al., "Opportunities to Grow Tribal Clean Energy in the US," *Science* 384, no. 6692 (2024): 163–165.

120. Chandra Farley et al., "Advancing Equity in Utility Regulation" (Berkeley, CA: Grid Modernization Laboratory Consortium, US Department of Energy, November 2, 2021).

121. Nik Heynen, "Bending the Bars of Empire from Every Ghetto for Survival: The Black Panther Party's Radical Antihunger Politics of Social Reproduction and Scale," *Annals of the Association of American Geographers* 99, no. 2 (2009): 406–422.

122. Angel, "Irregular Connections"; also Becker, Angel, and Naumann, "Energy Democracy as the Right to the City."

123. "Principles of Environmental Justice," *Race, Poverty & the Environment* 21, no. 2 (2017): 82–83.

124. My analysis of democracy is based on Ranciere's theorization of politics and radical democracy as well as organic democratic organizing to contest racial apartheid in the Black freedom movement. Jacques Ranciere, *Disagreement: Politics and Philosophy*, (Minneapolis, MN: University of Minnesota Press, 2004); Barbara Ransby, *Ella Baker and the Black Freedom Movement* (Chapel Hill, NC: The University of North Carolina Press, 2003); Nikhil Pal Singh, *Black Is a Country: Race and the Unfinished Struggle for Democracy* (Cambridge, MA: Harvard University Press, 2004).

125. Sean Sweeney, Kylie Benton-Connell, and Lara Skinner, "Power to the People: Toward Democratic Control of Electricity Generation," Working Paper No. 4, Trade Unions for Energy Democracy (The Worker Institute, Cornell University ILR School and the Rosa Luxemburg Stiftung New York Office, June 2015), 23, http://unionsforenergydemocracy.org/wp-content/uploads/2015/06/TUED-Power-to-the-Peoplefinal.pdf.

126. Robert D. Bullard, *Dumping in Dixie: Race, Class, and Environmental Quality* (Boulder, CO: Westview Press, 2000); Hilda E. Kurtz, "Acknowledging the Racial State: An Agenda for Environmental Justice Research," *Antipode* 41, no. 4 (2009): 684–704; David N. Pellow, "Toward a Critical Environmental Justice Studies," *Du Bois Review: Social Science Research on Race* 13, no. 2 (2016): 221–236; on urban political ecologies, see William Cronon, *Nature's Metropolis: Nature's Metropolis: Chicago and the Great West* (New York: W.W. Norton & Company, Inc., 1992); Nik Heynen, Maria Kaika, and Erik Swyngedouw, *In the Nature of Cities: Urban Political Ecology and the Politics of Urban Metabolism* (London: Routledge, 2006); on nature, see Neil Smith, *Uneven Development: Nature, Capital, and the Production of Space* (London: Verso, 2010).

127. Dianne Rocheleau, Barbara Thomas-Slayter, and Esther Wangari, eds., *Feminist Political Ecology: Global Issues and Local Experiences* (London: Routledge, 1996); Sharlene Mollett and Caroline Faria, "Messing with Gender in Feminist Political Ecology," *Geoforum* 45 (2013): 116–125; Sapana Doshi, "Embodied Urban Political Ecology: Five Propositions," *Area* 49, no. 1 (2017): 125–128; Nik Heynen, "Urban Political Ecology III: The Feminist and Queer Century," *Progress in Human Geography* 42, no. 3 (2018): 446–452.

128. See for instance Katz, "Bad Elements"; Linda Peake et al., eds., *A Feminist Urban Theory for Our Time: Rethinking Social Reproduction and the Urban* (Hoboken, NJ: Wiley, 2021).

129. Ian Kim, Emily Kirsch, and Chris Reyes, "Making Green Work: Best Practices in Green-Collar Job Training" (Oakland, CA: Ella Baker Center for Human Rights, 2010), https://ellabakercenter.org/sites/default/files/downloads/making-green-work.pdf; Derrick Johnson and Ashura Lewis, "Organizing for Energy Democracy in Rural Electric Cooperatives," in *Energy Democracy: Advancing Equity in Clean Energy Solutions*, ed. Denise Fairchild and Al Weinrub (Washington, DC: Island Press, 2017), 93–112; Nathaniel Smith, "Equity Matters: On the Shoulders of Nameless Others," *The Atlanta Voice*, April 6, 2018, https://www.theatlantavoice.com/articles/on-the-shoulders-of-nameless-others/.

130. Ransby, *Ella Baker and the Black Freedom Movement*, 368, 269.

131. Fairchild and Weinrub, *Energy Democracy*, 6; Baker, *Revolutionary Power*; Lennon, "Energy Transitions in a Time of Intersecting Precarities."

132. See Becker, Angel, and Naumann, "Energy Democracy as the Right to the City"; Nik Heynen and Nikki Luke, "The Case for Reparations, Urban Political Ecology, and the Black Right to Urban Life," in *Turning up the Heat: Urban Political Ecology for a Climate Emergency*, ed. Maria Kaika et al. (Manchester: Manchester University Press, 2023), 129–142.

133. Christopher J. Manganiello, *Southern Water, Southern Power: How the Politics of Cheap Energy and Water Scarcity Shaped a Region* (Chapel Hill, NC: University of North Carolina Press, 2015); William D. Bryan, *The Price of Permanence: Nature and Business in the New South* (Athens: University of Georgia Press, 2018); Cater, *Regenerating Dixie*; Tyler McCreary and Richard Milligan, "The Limits of Liberal Recognition: Racial Capitalism, Settler Colonialism, and Environmental Governance in Vancouver and Atlanta," *Antipode* 53, no. 3 (2021): 724–744; Richard Milligan, Tyler McCreary, and Na'Taki Osborne Jelks, "Improvising against the Racial State in Atlanta: Reimagining Agency in Environmental Justice," *Environment and Planning C: Politics and Space* 39, no. 7 (2021): 1586–1605; Sellers, *Race and the Greening of Atlanta*; Kristin D. Phillips, "Southern Politics, Southern Power Prices: Race, Utility Regulation, and the Value of Energy," *Economic Anthropology* 10, no. 2 (2023): 197–212.

134. With one exception noted in the text, all interview participants are referred to by pseudonyms.

135. I did meet Southern Company and Alabama Power staff as part of a grant on the beneficial reuse of coal ash on which we were both contracted in January 2023.

136. Charles R. Hale, *Engaging Contradictions: Theory, Politics, and Methods of Activist Scholarship*, (Berkeley, CA: University of California Press, 2008); Kate Driscoll Derickson and Danny MacKinnon, "Toward an Interim Politics of Resourcefulness for the

Anthropocene," *Annals of the Association of American Geographers* 105, no. 2 (2015): 304–312; Kate Driscoll Derickson and Paul Routledge, "Resourcing Scholar-Activism: Collaboration, Transformation, and the Production of Knowledge," *The Professional Geographer* 67, no. 1 (2015): 1–7.

137. Richa Nagar et al., "Feminisms, Collaborations, Friendships: A Conversation," *Feminist Studies* 42, no. 2 (2016): 513; Richa Nagar, *Muddying the Waters: Coauthoring Feminisms Across Scholarship and Activism* (Urbana, IL: University of Illinois Press, 2014).

138. Katherine McKittrick, *Demonic Grounds: Black Women and the Cartographies of Struggle* (Minneapolis, MN: University of Minnesota Press, 2006), xxxi.

139. Cater, *Regenerating Dixie: Electric Energy and the Modern South.*

140. Stuart Hall and Doreen Massey, "Interpreting the Crisis," *Soundings* 44, no. 44 (March 25, 2010): 57–71.

Chapter 1

1. The name of the office that led this plan changed throughout the period discussed in this chapter. As of 2025, the Mayor's Office of Sustainability and Resilience leads energy policy for the City of Atlanta.

2. Sierra Club, "100% Commitments in Cities, Counties, & States," 2018, https://www.sierraclub.org/ready-for-100/commitments.

3. Joshua Long and Jennifer L Rice, "From Sustainable Urbanism to Climate Urbanism," *Urban Studies* 56, no. 6 (2019): 992–1008; Stokes, *Short Circuiting Policy.*

4. At the time of this writing, Georgia Power has no stated target to decarbonize. The state initiated its first climate action planning process in 2023 as a condition of receiving IRA funding. Emily Jones, "Georgia Is Making Its First-Ever Climate Change Plan," WABE, July 10, 2023, https://www.wabe.org/georgia-is-making-its-first-ever-climate-change-plan/.

5. These goals emerged from feedback from 3,500 people in town halls and online surveys. City of Atlanta Mayor's Office of Resilience, "Clean Energy Atlanta: A Vision for a 100% Clean Energy Future."

6. Participant observation, August 21, 2018.

7. City of Atlanta Mayor's Office of Resilience, "Clean Energy Atlanta: A Vision for a 100% Clean Energy Future," 6.

8. City of Atlanta Mayor's Office of Resilience, 6.

9. Charles Rutheiser, *Imagineering Atlanta* (London: Verso, 1996), 5.

10. Paul Gaston, *New South Creed.*

11. Stone asserts Atlanta governance emerges not from "the formal machinery of government" alone, "but rather the informal partnership between city hall and the downtown business elite." Stone, *Regime Politics*, 3.

12. Larry Keating, *Atlanta: Race, Class And Urban Expansion* (Philadelphia: Temple University Press, 2001), 2.

13. Mekko Chebon et al., "Eviction Notice from the Mvskoke People to Mayor Andre Dickens and Cop City," DEFEND THE ATLANTA FOREST, March 10, 2023, https://defendtheatlantaforest.org/2023/03/10/eviction-notice-from-the-mvskoke-people/; Roxanne Dunbar-Ortiz, An Indigenous People's History of the United States (Boston: Beacon Press, 2014), 111–114.

14. Rutheiser, *Imagineering Atlanta.*

15. Du Bois, *Black Reconstruction in America, 1860–1880,* 581.

16. Gaston, *New South Creed.*

17. LeeAnn Lands, *The Culture of Property: Race, Class, and Housing Landscapes in Atlanta, 1880–1950* (Athens, GA: University of Georgia Press, 2009), 19.

18. These same neighborhoods were first to receive water and sewage services that mitigated yellow fever and other diseases as well as street paving and sidewalks. James Michael Russell, *Atlanta, 1847–1890: City Building in the Old South and the New* (Baton Rouge, LA: Louisiana State University Press, 1988), 215–230.

19. Franklin M. Garrett, *Atlanta and Environs: A Chronicle of Its People and Events, 1880s–1930s*, vol. Volume II (Athens: University of Georgia Press, 1954), 309.

20. "A Great Exhibit, Which Demonstrates the Utility of the Mightiest Power in the World," *The Atlanta Journal*, October 26, 1895.

21. Wade Wright, *History of the Georgia Power Company, 1855–1956* (Atlanta, GA: Georgia Power Company, 1957), 21–22.

22. The city was part-owner of Atlanta Gas Light Company and paid its annual bills for lighting using stock dividends. In 1887, the city sold its shares to contribute to Georgia Technical College and build school and park facilities. See: Wright, 23, 2, 19.

23. Annual Reports of Officers of the City of Atlanta, for the Year Ending Dec. 31 1888, pp. 213–214. Quoted in Garrett, *Atlanta and Environs*, Volume II: 181–182.

24. Murray Melbin, "Night as Frontier," *American Sociological Review* 43, no. 1 (1978): 3.

25. "Should the City Own the Light Plant? Prominent Citizens Talk," *The Atlanta Constitution*, October 17, 1897.

26. Wright, *History of the Georgia Power Company, 1855–1956*, 42–45.

27. Mary Beth Reed et al., "Historic Streetcar Systems in Georgia," 2012, 42, http://www.dot.ga.gov/InvestSmart/Environment/CulturalResources/Pubs/GAStreetcar.pdf.

28. Wright, *History of the Georgia Power Company, 1855–1956*, 11.

29. Don L. Klima, "Breaking Out: Streetcars and Suburban Development, 1872–1900," *The Atlanta Historical Journal* 26, no. 2–3 (1982): 67–82.

30. Timothy J. Crimmins, "Bungalow Suburbs: East and West," *The Atlanta Historical Journal* 26, no. 2–3 (1982): 83–94.

31. "St. Railway Franchises Intended as a Whip," *The Constitution*, August 16, 1899.

32. Garrett, *Atlanta and Environs*, Volume II, 428.

33. Rutheiser, *Imagineering Atlanta*.

34. Lands, *The Culture of Property*.

35. On nightlife in turn-of-the-century Atlanta, Gretchen Ehrmann Maclachlan, "Women's Work: Atlanta's Industrialization and Urbanization, 1879–1929" (PhD, Atlanta, Georgia, Emory University, 1992).

36. Ray Stannard Baker, *Following the Color Line: An Account of Negro Citizenship in the American Democracy* (New York: Doubleday, Page & Company, 1908), 30.

37. Tera W. Hunter, *To 'joy My Freedom: Southern Black Women's Lives and Labors After the Civil War* (Cambridge, MA: Harvard University Press, 1997), 99.

38. Gregory Mixon, *The Atlanta Riot: Race, Class, and Violence in a New South City* (Gainesville, FL: University Press of Florida, 2005); Reed et al., "Historic Streetcar Systems in Georgia."

39. Hunter, *To 'joy My Freedom*.

40. "Alderman Key's Views," *The Atlanta Constitution*, August 26, 1906, also Mixon, *The Atlanta Riot*, 56–59. The *Atlanta Journal of Labor* also militated for municipal ownership, Cater, *Regenerating Dixie*, 48.

41. Tom Baxter, "Streetcar Riots: Old Times Must Not Be Forgotten," *The Atlanta Journal-Constitution*, September 26, 1996.

42. David Fort Godshalk, *Veiled Visions: The 1906 Atlanta Race Riot and the Reshaping of American Race Relations* (Chapel Hill, NC: The University of North Carolina Press, 2005).

43. Godshalk contends streetcar motormen and the company were complicit in the devastation. Godshalk, 91; Baker, *Following the Color Line: An Account of Negro Citizenship in the American Democracy*, 26–35.

44. "Street Railways Under Commission," *The Atlanta Georgian and News*, July 20, 1907, Georgia Historic Newspapers, https://gahistoricnewspapers.galileo.usg.edu/lccn/sn89053728/1907-07-20/ed-1/seq-1/.

45. Beder, *Power Play*.

46. Cater, *Regenerating Dixie*, 51; "The Great and Historic Legislature of 1907," *The Atlanta Georgian and News*, August 19, 1907, Georgia Historic Newspapers.

47. Cater, *Regenerating Dixie*, 68; Bernard Suttler, "The Water Power Trust," *The Atlantian*, December 1, 1911.

48. "Claims Power Co. Is Far from Ruin," *The Atlanta Constitution*, May 21, 1918.

49. "Gas and Light Increases Awarded," *The Constitution*, August 18, 1918; also Wright, *History of the Georgia Power Company, 1855–1956*, 169–170.

50. "Gas and Light Increases Awarded."

51. "Full Text of Supreme Court Car Fare Decision," *The Atlanta Constitution*, March 16, 1919.

52. "Municipal Ownership for Atlanta Defeated in House Committee," *The Atlanta Constitution*, July 24, 1919.

53. Wright, *History of the Georgia Power Company, 1855–1956*, 262.

54. "Council Moves to Get TVA Power for City; Savage Named to Investigate Feasibility," *The Atlanta Constitution*, August 7, 1934.

55. Taft and Heys, *Big Bets*.

56. Each state utility remains responsive to its state public service commission. See: Taft and Heys.

57. Thomas Westbury, "Power to the People," *The Atlanta Constitution*, October 28, 1979, Sunday edition; Georgia Public Service Commission, "Electric," 2020, https://psc.ga.gov/utilities/electric/.

58. Baker, *Following the Color Line*; Clifford M. Kuhn, Harlon E. Joye, and E. Bernard West, *Living in Atlanta: An Oral History of the City, 1914–1948* (Atlanta and Athens, GA: Atlanta Historical Society and The University of Georgia Press, 1990); Godshalk, *Veiled Visions*.

59. Kuhn, Joye, and West, *Living in Atlanta*, 37; also Kevin Kruse, *White Flight*; Lands, *The Culture of Property*.

60. Rodriguez, *Diverging Space for Deviants*, 41; on zoning, see also Ronald H. Bayor, *Race and the Shaping of Twentieth-Century Atlanta* (Chapel Hill, NC: University of North Carolina Press, 1996), 55.

61. Nye, *Electrifying America*, 54.

62. Hunter, *To 'joy My Freedom*, 146–147.

63. Hunter, 152; Garrett, *Atlanta and Environs*, Volume II, 607–609. Kuhn, Joye, and West, *Living in Atlanta*.

64. Klima, "Breaking Out: Streetcars and Suburban Development, 1872–1900."

65. Wright, *History of the Georgia Power Company, 1855–1956*, 152.

66. *Addresses by Preston S. Arkwright President of the Georgia Railway & Power Company* (Georgia Railway and Power Company, 1926), 103–104.

67. Conor Harrison, "Extending the 'White Way': Municipal Streetlighting and Race, 1900–1930," *Social & Cultural Geography* 16, no. 8 (2015): 950–973; on disparities in public services in Atlanta, see Jay Winston Driskell Jr., *Schooling Jim Crow: The Fight for Atlanta's Booker T. Washington High School and the Roots of Black Protest Politics* (Charlottesville, VA: University of Virginia Press, 2014), 71.

68. Lands, *The Culture of Property*.

69. Bayor, *Race and the Shaping of Twentieth-Century Atlanta*, 55–60.

70. Hunter, *To 'joy My Freedom*, 100.

71. Rodriguez, *Diverging Space for Deviants*; on organizing against rate increases by the Atlanta Chapter of the Welfare Rights Organization see: "Public Service Commission—Hearing Reporter—Public Hearing Minutes—Minute Book 1971–1974," 1974, 108, Georgia State Archives, Vol2-8398, 017-04-026; Dennis Creech and Natalie Brown, "Energy Use and the Environment," in *Sprawl City: Race, Politics, and Planning in Atlanta*, ed. Robert Bullard, Glenn S. Johnson, and Angel O. Torres (Washington, DC: Island Press, 2000), 187–207.

72. Bayor, *Race and the Shaping of Twentieth-Century Atlanta*.

73. "Hearing of Car Fare Testimony Will End Today," *The Atlanta Constitution*, July 19, 1927.

74. "10-Cent Car Fare Ruling Accepted by Power Company," *The Atlanta Constitution*, December 7, 1927.

75. Kruse, *White Flight*, 111. In 1945, the Southern Regional Council further observed that "no outlying Negro sections are served by trackless trolleys and buses, nor are there any of these sections served by feeder bus lines. These services are made available to many, although not all, outlying white sections of the city." New Black suburbs in Dixie Hills and Pine Acres were denied service by Georgia Power. Bayor, *Race and the Shaping of Twentieth-Century Atlanta*, 189.

76. Kruse, *White Flight*, 111.

77. Kuhn, Joye, and West, *Living in Atlanta*, 120.

78. Alex Cherup, "Servicemen Slain on Streetcars by Motormen with Guns: A Legal History and Analysis of the Killings of Veterans Walter Lee Johnson and Madison Harris" (Northeastern University School of Law, Boston, MA: Civil Rights and Restorative Justice, 2014), https://repository.library.northeastern.edu/downloads/neu:m042hg03m?datastream_id=content.

79. James L. Delk to Georgia Power Company, July 24, 1948, MSS 619, Box 116, Folder 5, Atlanta History Center; Manager Transportation Department, Georgia Power Company to William P. Jones, April 23, 1946, MSS 619, Box 116, Folder 5, Atlanta History Center.

80. Tomiko Brown-Nagin, *Courage to Dissent: Atlanta and the Long History of the Civil Rights Movement* (Oxford, UK: Oxford University Press, 2011), 124. Transit companies only employed white drivers until the mid-1960s. Kuhn, Joye, and West, *Living in Atlanta, 82*.

81. W. Hubert Joiner, *Let Us Reason Together: A History of Labor Relations of the Georgia Power Company* (Atlanta, GA: Georgia Power Company, 1979), 134, 217.

82. Kruse, *White Flight,* 19.

83. Cobb, *The Selling of the South*, 144–145.

84. Stone, *Regime Politics*; Bayor, *Race and the Shaping of Twentieth-Century Atlanta*; Hobson, *The Legend of the Black Mecca*.

85. Kruse, *White Flight.*

86. Huber, *Lifeblood*, 18–19.

87. Bayor, *Race and the Shaping of Twentieth-Century Atlanta*; Sellers, *Race and the Greening of Atlanta;* Jason Henderson, "Secessionist Automobility: Racism, Anti-Urbanism, and the Politics of Automobility in Atlanta, Georgia," *International Journal of Urban and Regional Research* 30, no. 2 (2006): 293–307; Rob Gioielli, "Mobility, Race, and Climate in Postwar Atlanta," *American Studies* 60, no. 3–4 (2021).

88. US Department of Commerce Bureau of the Census, "Population of Georgia, By Counties, April 1, 1950, 1950 Census of Population Preliminary Counts," September 6, 1950, https://www2.census.gov/library/publications/decennial/1950/pc-02/pc-2-34.pdf; US Census Bureau, "Counties in South and West Lead Nation in Population Growth," The United States Census Bureau, April 18, 2019, https://www.census.gov/newsroom/press-releases/2019/estimates-county-metro.html.

89. "Most to Least Segregated Cities," Othering & Belonging Institute, 2024, https://belonging.berkeley.edu/most-least-segregated-cities.

90. This was a slogan of the Southern Company, Taft and Heys, *Big Bets*.

91. Cebul, "Creative Competition," 52.

92. "Georgia Power Company Yearbook" (Atlanta, GA, 1927), 35, MSS 363, Box 1, Atlanta History Center.

93. Cobb, *The Selling of the South.*

94. Cebul, "Creative Competition," 63.

95. For example recruitment pitch, see "Georgia Power Chiefs Preview Film on Poultry," *The Atlanta Constitution*, January 19, 1946.

96. Glenn E. McLaughlin and Stefan Hyman Robock, *Why Industry Moves South: A Study of Factors Influencing the Recent Location of Manufacturing Plants in the South* (Washington, DC: Committee of the South, National Planning Association, 1949), 71.

97. McLaughlin and Robock, 60.

98. Cobb, *The Selling of the South*, 217.

99. Moore, "An Evaluation of the Art and Techniques of Corporate Public Relations as Practices by the Georgia Power Company," 147–159.

100. Schulman, *From Cotton Belt to Sunbelt.*

101. Georgia Power Company, "Power Marches Ahead of War's Mounting Demands," *The Macon News*, October 6, 1943; "Macon Plant Adds Millions of Kilowatt Hours," *The Atlanta Constitution*, June 5, 1941; on the importance of the military industrial complex to Georgia's industrialization, see Schulman, *From Cotton Belt to Sunbelt.*

102. Malm, *Fossil Capital.*

103. Andrew Marshall Hamer, "Urban Perspective for the 1980s," in *Urban Atlanta: Redefining the Role of the City*, ed. Andrew Marshall Hamer (Atlanta, GA: Business Pub. Division, College of Business Administration, Georgia State University, 1980), 9; also Glenwood Ross, David L. Sjoquist, and Matthew Wooten, "Tracking the Economy of the City of Atlanta: Past Trends and Future Propsects," in *Past Trends and Future Prospects of the American City: The Dynamics of Atlanta*, ed. David L. Sjoquist (Lanham, MD: Lexington Books, 2009), 54.

104. Robert Bullard, Glenn S. Johnson, and Angel O. Torres, eds., *Sprawl City: Race, Politics, and Planning in Atlanta* (Washington, DC: Island Press, 2000). On Black working-class organizing, see Hobson, *The Legend of the Black Mecca*, 144.

105. Mary Mathewes Kassis and David Boldt, "Historical Perspective of Georgia's Economy, A" (Atlanta, GA: Andrew Young School of Public Policy, February 2005); US Bureau of Labor Statistics, "Atlanta Area Employment—October 2022," 2022, https://www.bls.gov/regions/southeast/news-release/areaemployment_atlanta.htm.

106. Ross, Sjoquist, and Wooten, "Tracking the Economy of the City of Atlanta: Past Trends and Future Propsects."

107. Immergluck, *Red Hot City* (Oakland, California: University of California Press, 2022), 56.

108. "Manufacturing Does Big Part in Development of State, Cites Arkwright," *The Macon News*, April 5, 1928.

109. Tammy Joyner, "Packing a New Kind of Punch," *The Atlanta Journal-Constitution*, April 19, 1999; also Marable, "Power to the People?" 13, 10.

110. See, for instance, analysis of the film industry tax credit developed by the Metro Atlanta Chamber of which Georgia Power is a member. Pendleton Group and Metro Atlanta Chamber, "Key Considerations in Analyzing Recent Tax Provision Audits," *Atlanta Journal-Constitution*, 2023, 6, https://acrobat.adobe.com/link/track?uri=urn%3Aaaid%3Ascds%3AUS%3Abaee3a7b-6328-35d1-a9df-24c90bbc761d&utm_source=Iterable&utm_medium=email&utm_campaign=campaign_6203774&viewer%21megaVerb=group-discover.

111. The pitch is no longer in a printed yearbook but located at selectgeorgia.com. Since 2020, Georgia Power has partnered with Visa, Microsoft, PepsiCo, Papa Johns, Sam's Club, Carvana, Remington Firearms, Asos, FanDuel, NewCold, Vanderlande, Deluxe, SatisPie, Ken's Foods, and Meissner Corporation among others. Georgia Power Community and Economic Development, "Georgia Overview," 2023, https://www.selectgeorgia.com/discover-georgia/georgia-overview/.

112. Interview with energy consultant, September 13, 2019.

113. Danny Kanso, "Georgia Tax Breaks Don't Deliver," *Georgia Budget and Policy Institute*, February 26, 2021, https://gbpi.org/georgia-tax-breaks-dont-deliver/.

114. Peter Brand, "Green Subjection: The Politics of Neoliberal Urban Environmental Management," *International Journal of Urban and Regional Research* 31, no. 3 (2007): 616–632; also Jennifer L. Rice, "Climate, Carbon, and Territory: Greenhouse Gas Mitigation in Seattle, Washington," *Annals of the Association of American Geographers* 100, no. 4 (2010): 929–937; Long and Rice, "From Sustainable Urbanism to Climate Urbanism."

115. Darryl Fears, "City Warms to Solar Energy Proposals," *The Atlanta Constitution*, April 20, 1995.

116. Dennis Creech, "Empowerment Zone Will Look to the Sun," *The Atlanta Constitution*, May 18, 1995.

117. Maria Saporta, "Atlanta Rejects Turner Funds for Energy Conservation Plan," *The Atlanta Constitution*, July 10, 1998.

118. The organizing committee and state legislature opposed city proposals to charge Olympic tourists fees on tickets or a temporary sales tax to capture some revenue from the events. Rutheiser, *Imagineering Atlanta*, 248.

119. Robert DiGiacomo, "Sprawling Atlanta Seeks New Routes to the Future," *Grist*, May 15, 2008, https://grist.org/article/atlanta2/; Michael Ulmer, "The Wisdom and Wit of Mayor Shirley Franklin," *Atlanta City Council*, March 28, 2022, https://atlantacitycouncil.medium.com/the-wisdom-and-wit-of-mayor-shirley-franklin-3c52b632010b; on the Beltline, see Dan Immergluck and Tharunya Balan, "Sustainable for Whom? Green Urban Development, Environmental Gentrification, and the Atlanta Beltline," *Urban Geography* 39, no. 4 (2018): 546–562.

120. Immergluck, *Red Hot City*, 59.

121. Interview with energy advocate, July 30, 2019

122. Interview with former city employee, June 25, 2019.

123. Interview with energy advocate, July 30, 2019; the difficulty in funding projects at the city came up in interviews with city employees, March 20, 2019; May 21, 2019.

124. City of Atlanta Mayor's Office of Sustainability, "Power to Change: Atlanta's Citywide Sustainability Initiative," 2014, http://uccrnna.org/wp-content/uploads/2017/06/91_Atlanta_2014_Power-to-Change-Plan.pdf.

125. Josh Green, "Atlanta Leaders Want to Power the City with 100 Percent Clean Energy by 2035. Can It Be Done?," *Atlanta Magazine*, April 19, 2018, https://www.atlantamagazine.com/news-culture-articles/atlanta-leaders-want-power-city-100-percent-clean-energy-2035-can-done/.

126. Interview with former city employee June 2, 2019.

127. City of Atlanta Mayor's Office of Resilience, "Clean Energy Atlanta: A Vision for a 100% Clean Energy Future," 42.

128. NREL, "Georgia Residential Energy Efficiency Potential," ResStock, 2017, https://resstock.nrel.gov/factsheets/GA.

129. Meris Lutz, "Knock, Knock. It's Atlanta. The City Wants to Lower Your Energy Bill.," *The Atlanta Journal-Constitution*, November 21, 2023, https://www.ajc.com/news/business/knock-knock-its-atlanta-the-city-wants-to-lower-your-energy-bill/4PS3DW7C6FAIPMSOGO6JXJXZMA/.

130. 2021 was the last year for which this study was available. Georgia Power spent 0.63 percent of revenue on energy efficiency compared to a nationwide average of 2.2 percent. Energy efficiency investments increased 15 percent in the 2022 IRP. Mike Specian et al., "2023 Utility Energy Efficiency Scorecard" (American Council for an Energy-Efficient Economy, August 2023), https://www.aceee.org/sites/default/files/pdfs/U2304.pdf.

131. Interview with former city employee June 25, 2019.

132. Interviews with former city employee June 25, 2019; with energy advocate, July 30, 2019; with sustainability staff August 23, 2019.

133. Marni von Wilpert, "City Governments Are Raising Standards for Working People—and State Legislators Are Lowering Them Back Down," *Economic Policy Institute*, August 26, 2017, https://www.epi.org/publication/city-governments-are-raising-standards-for-working-people-and-state-legislators-are-lowering-them-back-down/; Richard Shragger, "State Preemption of Local Laws: Preliminary Review of Substantive Areas" (Legal Effort to Address Preemption Project, May 2017), https://www.abetterbalance.org/wp-content/uploads/2017/08/State-Preemption-of-Local-Laws.pdf.

134. Interview with sustainability staff August 23, 2019.

135. Wilborn Nobles, "Atlanta Chief Sustainability Officer Working to Fill Departmental Vacancies," *The Atlanta Journal-Constitution*, December 2, 2022, https://www.ajc.com/news/atlanta-news/atlanta-working-to-create-mayors-office-of-sustainability-and-resilience/YDFC5VMGARGKNJ4L7WRFGUPB44/.

136. Eva Dickerson, "How the Black Misleadership Class Provides Cover to Cop City," *Hammer & Hope*, 2023, https://hammerandhope.org/article/andre-dickens-cop-city-black-politicians.

137. Keating, *Atlanta*, 6.

138. Meris Lutz, "Atlanta Inches Toward 100% Clean Energy Goal as City Girds for Trump," *The Atlanta Journal-Constitution*, January 13, 2025, https://www.ajc.com/news/business/how-close-is-atlanta-to-its-100-clean-energy-goal/JPZJ3IDISRF6TDSJDMC7CNG7E4/.

139. Immergluck, *Red Hot City*, 7.

140. Rodriguez, *Diverging Space for Deviants*, 6–7.

141. Nathan Rott and Molly Samuel, "'Tidal Wave' of Power Shut-Offs Looms as Nation Grapples with Heat," NPR, July 28, 2020, https://www.npr.org/2020/07/28/892995058/tidal-wave-of-power-shut-offs-looms-as-nation-grapples-with-heat. Georgia Power did offer a payment plan to allow customers to pay off accrued debt.

142. Exelon disconnected more people yet served many more customers than Georgia Power. Selah Goodson Bell et al., "Powerless in the United States: How Utilities Drive Shutoffs and Energy Injustice" (Center for Biological Diversity, Energy and Policy Institute, BailoutWatch, January 2023), https://www.biologicaldiversity.org/programs/energy-justice/pdfs/Powerless-in-the-US_Report.pdf.

143. Danielle Kurtzleben, "Job Losses Higher Among People of Color During Coronavirus Pandemic," *NPR*, April 22, 2020, https://www.npr.org/2020/04/22/840276956/minorities-often-work-these-jobs-they-were-among-first-to-go-in-coronavirus-layo.

144. *Rose v. Raffensperger* complaint (District Court for the Northern District of Georgia Atlanta Division July 14, 2020).

145. *Rose v. Raffensperger* complaint.

146. Phillips, "Southern Politics, Southern Power Prices," 202.

147. Mark Niesse and Drew Kann, "How a Rare Voting Rights Win Took Two Races off Georgia Ballots," *The Atlanta Journal-Constitution*, August 25, 2022.

148. Personal communication with energy advocate, June 6, 2025.

Chapter 2

1. *Deploy23 Plenary 1: Welcome to Deploy23 (Jigar Shah)* (Washington, DC, 2023), https://www.youtube.com/watch?v=y7trk2OzLDc.

2. Drew Kann, "Georgia Power, PSC Staff Reach Deal on Final Vogtle Cost to Customers," *The Atlanta Journal-Constitution*, August 30, 2023.

3. *Deploy23 Plenary 1: Countdown to Liftoff*, 2023, https://www.youtube.com/watch?v=4a1MA71ozS4.

4. Julie Kozeracki and et al, "Pathways to Commercial Liftoff: Advanced Nuclear" (US Department of Energy, 2023), https://liftoff.energy.gov/wp-content/uploads/2023/05/20230320-Liftoff-Advanced-Nuclear-vPUB-0329-Update.pdf.

5. Participant observation, September 27, 2023.

6. Conor Harrison, "Electricity Capital and Accumulation Strategies in the US Electricity System."

7. Harrison, "Electricity Capital and Accumulation Strategies in the US Electricity System."

8. Neferti X. M. Tadiar, "Life-Times of Disposability within Global Neoliberalism," *Social Text* 31, no. 2 (2013): 26.

9. Conor Harrison, "The Historical–Geographical Construction of Power: Electricity in Eastern North Carolina," *Local Environment* 18, no. 4 (2013): 469–86. Geographers have studied how people embody financialization, including household mortgages and austerity imposed to manage city and state debt. See: Melissa García-Lamarca and Maria Kaika, "'Mortgaged Lives': The Biopolitics of Debt and Housing Financialisation," *Transactions of the Institute of British Geographers* 41, no. 3 (2016): 313–327; Jamie Peck and Heather Whiteside, "Financializing Detroit," *Economic Geography* 92, no. 3 (2016): 235–268; Laura Pulido, "Flint, Environmental Racism, and Racial Capitalism," *Capitalism Nature Socialism* 27, no. 3 (2016): 1–16.

10. Marcus Franklin et al., "Lights Out in the Cold: Reforming Utility Shut-Off Policies as If Human Rights Matter" (NAACP Environmental and Climate Justice Program, March 2017), http://www.naacp.org/wp-content/uploads/2017/12/Lights-Out-in-the-Cold_NAACP.pdf; Michelle Graff et al., "Which Households Are Energy Insecure? An Empirical Analysis of Race, Housing Conditions, and Energy Burdens in the United States," *Energy Research & Social Science* 79 (2021): 102144.

11. Georgia Public Service Commission, "The PSC: An Introduction to Your Georgia Public Service Commission," 2023, https://psc.ga.gov/about-the-psc/.

12. Georgia Public Service Commission, "Home," Georgia Public Service Commission, 2020, https://psc.ga.gov/.

13. Tadiar, *Remaindered Life*, 71.

14. Austin Abby et al., "Utilities' Low-Income Discount Programs Help Address Energy Insecurity, but Some US States Lag Behind," Center on Global Energy Policy at Columbia University SIPA, September 18, 2024, https://www.energypolicy.columbia.edu/utilities-low-income-discount-programs-help-address-energy-insecurity-but-some-us-states-lag-behind/.

15. Dominic J. Bednar and Tony G. Reames, "Recognition of and Response to Energy Poverty in the United States," *Nature Energy* 5, no. 6 (2020): 432–439; Drew Kann, "Georgia Power, PSC Staff Reach Deal on Final Vogtle Cost to Customers," *The Atlanta Journal-Constitution*, August 30, 2023.

16. Diana Hernández and Jennifer Laird, "Surviving a Shut-Off: US Households at Greatest Risk of Utility Disconnections and How They Cope," *American Behavioral Scientist* 66, no. 7 (2022): 856–880; US Energy Information Administration, "In 2020, 27% of US Households Had Difficulty Meeting Their Energy Needs," April 11, 2022, https://www.eia.gov/todayinenergy/detail.php?id=51979.

17. Maria Kaika, "Interrogating the Geographies of the Familiar: Domesticating Nature and Constructing the Autonomy of the Modern Home," *International Journal of Urban and Regional Research* 28, no. 2 (2004): 270.

18. Taft and Heys, *Big Bets*, 62.

19. Lowe, *The Intimacies of Four Continents*, 21.

20. Cowan, *More Work for Mother*, 13.

21. Wright, *History of the Georgia Power Company, 1855–1956*, 93.

22. Deborah G. Felder, *A Century of Women: The Most Influential Events in Twentieth-Century Women's History* (Secaucus, NJ: Carol Publishing Group, 1999), 14.

23. Bob Johnson, "Energy Slaves: Carbon Technologies, Climate Change, and the Stratified History of the Fossil Economy," *American Quarterly* 68, no. 4 (2016): 955–979.

24. Georgia Power Company, "'These Are Ways I Work for You' Says Reddy Kilowatt," *The Atlanta Journal*, May 6, 1949.

25. Georgia Power Company, "Putting Electricity to Work on Your Farm," n.d., MS 363, Box 3, Folder 4, Kenan History Center.

26. Georgia Power Company, "Along Came Youth," *The Southern Israelite*, September 1933.

27. Postone, *Time, Labor, and Social Domination*, 214–15; on the industrialization of household labor, see Davis, *Women, Race & Class*; Cowan, *More Work for Mother*; Strasser, *Never Done*; Claudia Goldin, "The Quiet Revolution That Transformed Women's Employment, Education, and Family," *American Economic Review* 96, no. 2 (2006): 1–21.

28. Moore, "An Evaluation of the Art and Techniques of Corporate Public Relations as Practices by the Georgia Power Company," 67.

29. Felder, *A Century of Women*, 15.

30. From a 1935 edition of Georgia Power Snap Shots, quoted in Cebul, "Creative Competition," 61.

31. With increasing mechanization, the number of people employed in laundries began falling in the 1910s. Maclachlan, "Women's Work," 242; Tera W. Hunter, "'The Women Are Asking for BREAD, Why Give Them STONE?': Women, Work, and Protests in Atlanta and Norfolk During World War I," in *Labor in the Modern South*, ed. Glenn T. Eskew (Athens, GA: University of Georgia Press, 2001), 102–134.

32. See: Felder, *A Century of Women*; Hunter, *To 'joy My Freedom*; Kuhn, Joye, and West, *Living in Atlanta*. Dorothy Bolden, who founded the National Domestic Worker's Union of America in Atlanta in 1969, defined the devaluation of Black women's work as directly enriching white households because "we was helping them to create their money. We was saving them. They was living in luxury. We wasn't." Rosalind Bentley, "Dorothy Bolden: Helped Black Domestic Workers Harness Political Power," *Atlanta Journal-Constitution*, February 20, 2019, https://www.ajc.com/lifestyles/dorothy-bolden-helped-black-domestic-workers-harness-political-power/peRGIyDV2cjjP7uSHgIUuN/.

33. Cowan, *More Work for Mother*.

34. The first public housing project was constructed in Atlanta in 1936 and provided many tenants "their first indoor plumbing, steam heat and electricity." Kuhn, Joye, and West, *Living in Atlanta*, 51.

35. Needham, *Power Lines: Phoenix and the Making of the Modern Southwest*, 78–79; Ronald Tobey, *Technology as Freedom* (Berkeley, CA: University of California Press, 1996), chapter 5.

36. Mock, "The Electric Home and Farm Authority, 'Model T Appliances,' and the Modernization of the Home Kitchen in the South," 75–76.

37. Cebul, "Creative Competition," 59.

38. In 1960, 12.3 percent of all households in Georgia had AC in at least part of their households, compared to 1.9 percent of Black households. In 1980, 66 percent of total Georgia households had AC compared to 42.7 percent of Black households. Approximately 98 percent of all homes have air conditioning today. See Raymond Arsenault, "The End of the Long Hot Summer: The Air Conditioner and Southern Culture," *The*

Journal of Southern History 50, no. 4 (1984): 611; US Census Bureau, "2021 Atlanta—Heating, Air Conditioning, and Appliances—All Occupied Units," 2021, https://www.census.gov/programs-surveys/ahs/data/interactive/ahstablecreator.html?s_areas=12060&s_year=2021&s_tablename=TABLE3&s_bygroup1=1&s_bygroup2=1&s_filtergroup1=1&s_filtergroup2=1. Today, heating and cooling homes to a safe and comfortable temperature remains cost-prohibitive to many, especially in majority Black census tracts. See: Luling Huang et al., "Inequalities across Cooling and Heating in Households: Energy Equity Gaps," *Energy Policy* 182 (2023): 113748; also Tony Gerard Reames, "Targeting Energy Justice: Exploring Spatial, Racial/Ethnic and Socioeconomic Disparities in Urban Residential Heating Energy Efficiency," *Energy Policy* 97 (2016): 549–558.

39. Data compiled from Wright, *History of the Georgia Power Company, 1855–1956;* Georgia Power Company Annual Reports in the Kenan Research Center, Atlanta History Center archives, and Georgia Power Company Facts and Figures. Gaps represent years for which data is not available. Sales for 1900 are estimated.

40. Hirsh, *Power Loss*, 46.

41. Taft and Heys, *Big Bets*, 276, 249.

42. The rate of return is intended to attract additional capital for continued investment without accruing monopoly profits to the utility. Bonbright, *Principles of Public Utility Rates*; Kihm et al., "You Get What You Pay For: Moving Toward Value in Utility Compensation, Part 1, Revenue and Profit."

43. Georgia Power, "Facts & Figures."

44. Tom Greene, "Power Firm Granted 2 Per Cent Increase," *The Atlanta Journal*, December 12, 1968.

45. On politicization, see Erik Swyngedouw, "The Antinomies of the Postpolitical City: In Search of a Democratic Politics of Environmental Production," *International Journal of Urban and Regional Research* 33, no. 3 (2009): 601–620; Derickson and MacKinnon, "Toward an Interim Politics of Resourcefulness for the Anthropocene."

46. Katherine Perrotta, "Ruffled Feathers: The Great Speckled Bird as a Record of Student and Youth Activism in Atlanta, Georgia and the Southeast, 1968–1976," *American Educational History Journal* 45, no. 1/2 (January 2018): 39–54.

47. Tom Coffin, "Ga. Power," *Great Speckled Bird*, 1968, 8.

48. Georgia Power Project, "Georgia Power," *Great Speckled Bird*, 1972.

49. "Georgia Power Project," *Great Speckled Bird*, September 11, 1972, 15; Gene Guerrero, "Ga. Power Rate Hike?," *Great Speckled Bird*, August 28, 1972.

50. Bob Hall and Institute for Southern Studies, "Georgia Power Company: A Continuing Saga!!!," *Great Speckled Bird*, July 24, 1972, 23.

51. Joseph Hughes, "Case Study: Taking Back Power," *Southern Exposure*, 1973, 58.

52. Macon Legal Aid filed on behalf of Mrs. Ethel Matthews, Mrs. Pinky Stinson, the Atlanta Chapter of the Welfare Rights Organization, and Mrs. Dorothy Bolden of the Atlanta Domestic Worker's Union. "Public Service Commission—Hearing Reporter—Public Hearing Minutes—Minute Book 1971–1974," 108.

53. Interview with environmental advocate, May 25, 2023; also Jon Jacobs, "Greedy Georgia," *Great Speckled Bird*, July 12, 1971.

54. Maurice Fliess, "McDonald Aim: Marxist Purge," *The Atlanta Journal*, March 13, 1975.

55. Ron Taylor, "Rate Hike Ads Hit by FCC," *The Atlanta Journal*, December 10, 1973; Mike Berman, "Power Rate Hike Hit Again," *The Atlanta Constitution*, November 15, 1974; "Hearing Requested Against Bond Plan," *The Atlanta Journal*, January 11, 1974.

56. James Hightower, "New Concern for Utility: Rate Structure," *The Atlanta Constitution*, August 26, 1973; Sam Hopkins, "Ga. Power Hits Back at Carter," *The Atlanta Constitution*, December 14, 1974.

57. Georgia Power Project, "Georgia Power"; Bill Jordan, "Ga. Power Plugs for Rise," *The Atlanta Constitution*, August 29, 1972.

58. Jim Gray, "Ga. Power Suits Dismissed," *The Atlanta Constitution*, August 20, 1975; John Huey and Claudia Townsend, "Power Group Rips Rate Hike," *The Atlanta Constitution*, January 30, 1975. Notably, Georgia Power sued after the Commission refused to provide their full rate request as a taking of its right to earn a return on its invested capital, while customers were disallowed form suing if rates were too high. "Ga. Power Seeks Securities Sales to Finance Building," *The Atlanta Constitution*, March 1, 1973, 11-D; Sharon Bailey, "Georgia Power Sued by Group," *The Atlanta Constitution*, February 8, 1975; "Hearing Requested Against Bond Plan."

59. Jon Jacobs, "Bird Sues Georgia Power," *Great Speckled Bird*, September 9, 1974, 1.

60. Interview with environmental advocate, May 19, 2023.

61. Roger Friedman and Andrew Reisinger, Roger Friedman oral history interview, 2016-07-14, July 14, 2016, Great Speckled Bird Oral History Project, Social Change Collection. Special Collections and Archives, Georgia State University, https://digitalcollections.library.gsu.edu/digital/collection/GSB/id/10233/rec/18.

62. Bill Jordan, "Georgia Power Rate Hike Is in the Cards," *The Atlanta Constitution*, October 6, 1972.

63. Friedman and Reisinger, Roger Friedman oral history interview, July 14, 2016.

64. Interview with environmental advocate, May 25, 2023.

65. Hopkins, "Ga. Power Hits Back at Carter," 13-A.

66. Friedman and Reisinger, Roger Friedman oral history interview, July 14, 2016.

67. Hopkins, "Ga. Power Hits Back at Carter," 13-A.

68. Arnold Fitzgerald, "Georgia Power Rate Increase Under Fire," *Atlanta Voice*, June 23, 1973.

69. "A Special Supplement: Southern Power Companies," *Southern Exposure*, 1973, 53.

70. Marable, "Power to the People?" 10.

71. Tom Henderson, "Southern Co. Sells $152.9 Million in Common Stock," *The Atlanta Constitution*, September 19, 1974; *Big Bets (Full Program)* (Atlanta, GA: Mountain View Group, 2014), https://vimeo.com/97226639. This is ultimately what the Georgia Power Project had argued was necessary, not ratepayers funding construction costs. See Friedman and Reisinger, Roger Friedman oral history interview, July 14, 2016.

72. *Big Bets (Full Program)*.

73. Hall and Institute for Southern Studies, "Georgia Power Company: A Continuing Saga!!!," 23.

74. Jon Jacobs, "The City Should Buy Out Georgia Power," *Great Speckled Bird*, June 10, 1974; The Georgia Power Project, "A City-Owned Power Company?," *The Atlanta Voice*, March 23, 1974.

75. Joseph D. Kearney and Thomas W. Merrill, "The Great Transformation of Regulated Industries Law," *Columbia Law Review* 98, no. 6 (1998): 1323–1409.

76. Georgia Public Service Commission, "Electric."

77. Andrea Jones, "Ga. Power Ongoing," *Great Speckled Bird*, February 19, 1973, 8; Mike Raffauf, "Electric Service Bill," *Great Speckled Bird*, February 1, 1973.

78. Taft and Heys, *Big Bets*, 261.

79. Taft and Heys, *Big Bets*, 260.

80. Cater, *Regenerating Dixie*, 8.

81. Taft and Heys, *Big Bets*, 276; Beau Cutts, "Light Bills Won't Be Cut," *The Atlanta Constitution*, January 8, 1975.

82. Margaret Shannon, "Consumers Get a Utility Watcher," *The Atlanta Constitution*, November 30, 1975.

83. David Markiewicz, "Consumers Lack Voice at PSC as Big Utility Cases Loom, Advocates Say," *The Atlanta Journal-Constitution*, February 20, 2010, https://www.ajc.com/news/local/consumers-lack-voice-psc-big-utility-cases-loom-advocates-say/oHklWe6eQXk4r8BtwGMwPJ/; interview with consumer advocate, December 3, 2020

84. The Commission permitted the utility to add some construction costs to its rate base in the 1970s. Bob Deans, "Paying for Plant Vogtle to Be a Costly Matter for Georgia," *The Atlanta Constitution*, February 3, 1985, 6D.

85. See Neferti Tadiar on national debt that cast "future life-times of surplus labor and money (in the form of taxes [or rates])—value—that can be advanced, that is, cashed in now." Tadiar, "Life-Times of Disposability Within Global Neoliberalism," 26.

86. Educational Campaign for a Prosperous Georgia, "Legislature Looks at Plant Vogtle," *Great Speckled Bird*, 1985, 9; Neill Herring, "Vogtle Follies," *Great Speckled Bird*, September 1984; Neill Herring, "Georgia Power Sails into Stormy Weather," *Great Speckled Bird*, 1984.

87. Brian O'Shea, "Plan to Let Utilities Charge for Unfinished Plants Shelved," *The Atlanta Constitution*, November 20, 1987.

88. Hirsh, *Power Loss*; Beder, *Power Play*.

89. Industrial customers criticized the Commission for "leaving $200–300 million a year on the table" in negotiations with the utility that came out of the pockets of customers. Peter Mantius, "Georgia Power Agrees to Lower Rates," *The Atlanta Constitution*, December 19, 1998, D1.

90. *Big Bets (Full Program)*.

91. Harrison and Welton, "The States That Opted Out."

92. *Big Bets (Full Program)*.

93. *Big Bets (Full Program)*.

94. The partial meltdown of the Fukishima Daichi reactor in 2011 halted other new nuclear projects. Southern Company, "Plant Vogtle 3 and 4," 2020, http://www.southerncompany.com/innovation/nuclear-energy/plant-vogtle-3-and-4.html. LPO provided $3.8 billion in additional loan authority under the Trump administration. Darrell Proctor, "Vogtle Loan Guarantees Praised, Criticized," *POWER Magazine*, March 23, 2019, https://www.powermag.com/vogtle-loan-guarantees-praised-criticized/.

95. Tom Crawford, "The Bombshell of the Session," *The Press Sentinel*, February 11, 2009.

96. Stan Wise and Angela Spier Phelps, "Do Georgians Gain from PSC Vote on Nuke Plant?" *The Atlanta Constitution*, March 26, 2009. A14.

97. "Timeline: Plant Vogtle Through the Years," *Atlanta Journal-Constitution*, December 20, 2017, https://www.ajc.com/news/local-govt--politics/plant-vogtle-timeline/wfQKABLCNutdld3RQPNwaL/.

98. Kristi E. Swartz, "New Reactors on the Verge of Ok," *The Atlanta Journal-Constitution*, January 29, 2012, A1.

99. Nuclear Watch South, "Georgia General Assembly Votes to Sunset CWIP," March 27, 2018, https://www.nonukesyall.org/stop_CWIP.html#.

100. Interview with environmental justice organizer, May 23, 2019.

101. Tadiar, "Life-Times of Disposability Within Global Neoliberalism," 30.

102. Interview with energy consultant, September 13, 2019.

103. Interview with energy consultant, September 13, 2019.

104. Georgia residents paid the seventh highest electricity bills of any state in the US in 2022, and the seventeenth highest in 2023. US Energy Information Administration, "2022 Average Monthly Bill- Residential," accessed May 29, 2022, https://www.eia.gov/electricity/sales_revenue_price/pdf/table_5A.pdf; US Energy Information Administration, "2023 Average Monthly Bill- Residential," accessed January 10, 2025, https://www.eia.gov/electricity/sales_revenue_price/pdf/table_5A.pdf.

105. Drehobl, Ross, and Ayala, "How High Are Household Energy Burdens? An Assessment of National and Metropolitan Energy Burden Across the United States."

106. On VC Summer, Shelley Welton and Conor Harrison, "Lessons in Climate Derisking: The United States' Failed Nuclear Renaissance," *University of Pennsylvania Law Review*, 2025.

107. Bill Edge, "Commission Approves Continued Construction of Nuclear Power Plant Vogtle Project; Reduces Company Revenue Requirement by $1.7 Billion," December 21, 2017, http://www.psc.state.ga.us/GetNewsRecordAttachment.aspx?ID=752.

108. Edge.

109. Commissioner Bubba McDonald quoted in Bill Edge, "Georgia Public Service Commission Reduces Nuclear Construction Cost Recovery Fee," March 30, 2018, http://www.psc.state.ga.us/GetNewsRecordAttachment.aspx?ID=780.

110. Meg Mirshak, "Energy Secretary Moniz Visits Vogtle to Finalize Loan Guarantee," *The Augusta Chronicle*, February 20, 2014, https://www.augustachronicle.com/article/20140220/NEWS/302209860.

111. Darrell Proctor, "Vogtle Loan Guarantees Praised, Criticized."

112. Kann, "Georgia Power, PSC Staff Reach Deal on Final Vogtle Cost to Customers."

113. Anastaciah Ondieki, "High Energy Bills Burden Atlanta's Low-Income Residents," *The Atlanta Journal-Constitution*, August 1, 2018.

114. Mike Morris and Jack Spink, "Cold Metro Residents Apply for Energy Funds," *The Atlanta Journal-Constitution*, December 4, 2010.

115. Sarah Davis, "Atlantans Feel Pinch of Extreme Heat," *The Atlanta Journal-Constitution*, July 30, 2023, https://www.newspapers.com/image/988505820/.

116. Interview with energy advocate, July 30, 2019.

117. Phillips, "Southern Politics, Southern Power Prices," 202.

118. Matt Kempner, "Georgia Power Bills Could Rise: $2.2 Billion Rate Increase Proposed," *Atlanta Journal Constitution*, June 28, 2019, https://www.ajc.com/news/local/georgia-power-bills-could-rise-billion-rate-increase-proposed/RbrV522vtw0236tn2swIYO/.

119. Partnership for Southern Equity, "Fight the Hike," 2019, http://fightthehike.org/.

120. Participant Observation, October 17, 2019.

121. "Docket No. 42516: Georgia Power Company's 2019 Rate Case" (Atlanta, GA, September 30, 2019), https://www.youtube.com/watch?v=aPquuso44kA&t=1531s.

122. *Docket 42516—11/04/19*, 2019, https://www.youtube.com/watch?v=biJN45hB4c4.

123. "Docket No. 42516: Georgia Power Company's 2019 Rate Case," September 30, 2019.

124. "Grady baby" refers to people born in Atlanta's Grady Hospital. "Docket No. 42516: Georgia Power Company's 2019 Rate Case."

125. "Docket No. 42516: Georgia Power Company's 2019 Rate Case."

126. *Docket 42516—11/04/19*.

127. *Docket 42516—11/04/19*.

128. On the extraction of value from territories see: Gilmore, "Abolition Geography and the Problem of Innocence."

129. Dave Williams, "Georgia Public Service Commission Set to Vote on Georgia Power Rate Hike," *Gwinnett Daily Post*, December 15, 2019, https://www.gwinnettdailypost.com/curated/georgia-public-service-commission-set-to-vote-on-georgia-power/article_ce020432-1fb1-11ea-b69d-438013c38878.html.

130. While the Public Interest Advocacy staff represent residential, small commercial, and industrial customers, they work for the Commission and are not seen as having the independence of the since-disbanded Consumer's Utility Counsel.

131. "Docket No. 42516: Georgia Power Company's 2019 Rate Case. Consideration of Commissioner Advisory Staff's Recommendation," December 12, 2019, https://www.youtube.com/watch?v=cYpJ5Hsq_pM.

132. "Docket No. 42516: Georgia Power Company's 2019 Rate Case. Consideration of Commission Advisory Staff's Recommendation."

133. "Docket No. 42516: Georgia Power Company's 2019 Rate Case. Consideration of Commission Advisory Staff's Recommendation."

134. Nathaniel Smith, "The Fight Continues . . . ," December 19, 2019.

135. John Weaver, "Georgia Power Has 5,000 Residential Solar Net Metering Contracts and It's First Come, First Served," *pv magazine USA*, December 18, 2019, https://pv-magazine-usa.com/2019/12/18/georgia-power-has-5000-residential-solar-power-net-metering-contracts-first-come-first-serve/.

136. "Docket No. 42516: Georgia Power Company's 2019 Rate Case," December 17, 2019, https://www.youtube.com/watch?v=2L9x-Rg5UYY&t=0s.

137. Weaver, "Georgia Power Has 5,000 Residential Solar Net Metering Contracts and It's First Come, First Served."

138. Michael Wald, "Treat Southern Company Like a Government-Insured Bond," Seeking Alpha, December 18, 2019, https://seekingalpha.com/article/4313082-treat-southern-company-like-government-insured-bond.

139. Data from: Weaver, "Georgia Power Has 5,000 Residential Solar Net Metering Contracts and It's First Come, First Served"; Drew Kann, "Georgia Power's Rates Will Go up Again Next Month. Here's How Much," *The Atlanta Journal-Constitution*, December 17, 2024.

140. Dave Williams, "State Energy Regulators Approve Georgia Power Rate Hike," *Georgia Public Broadcasting*, December 21, 2022, https://www.gpb.org/news/2022/12/21/state-energy-regulators-approve-georgia-power-rate-hike; Kann, "Georgia Power, PSC Staff Reach Deal on Final Vogtle Cost to Customers."

141. Drew Kann, "Kemp Touts Deal to Temporarily Freeze Georgia Power Rates," *The Atlanta Journal-Constitution*, May 21, 2025, https://www.ajc.com/business/2025/05/kemp-touts-deal-to-temporarily-freeze-georgia-power-rates/.

Chapter 3

1. "Utility Company's Meeting Picketed," *Sun Herald*, May 22, 1985, A9.

2. Southern Company, "Planning for a Low-Carbon Future," April 2018, https://www.southerncompany.com/content/dam/southern-company/pdf/corpresponsibility/Planning-for-a-low-carbon-future.pdf; Sierra Club Georgia Chapter, "Southern Company Shareholder Activism," 2017, https://www.sierraclub.org/georgia/cefa/southernco; on the limits of shareholder activism, see Sarah Knuth, "Green Devaluation: Disruption, Divestment, and Decommodification for a Green Economy," *Capitalism Nature Socialism* 28, no. 1 (2017): 98–117.

3. Hurricane Michael was the first major hurricane to directly impact Georgia since the 1890s. John Schwartz, "Why Hurricane Michael's Power Caught Forecasters Off Guard," *The New York Times*, October 11, 2018, https://www.nytimes.com/2018/10/11/climate/hurricane-michael-science.html.

4. Participant observation, May 22, 2019.

5. Dipesh Chakrabarty, "Universalism and Belonging in the Logic of Capital," *Public Culture* 12, no. 2 (2000): 670.

6. Rob Nixon, *Slow Violence and the Environmentalism of the Poor*, 8.

7. Sarah Sharma, *In the Meantime: Temporality and Cultural Politics* (Durham, NC: Duke University Press Books, 2014). Postone argues that the linear, historical arc of progress follows technological change, while also naturalizing and reproducing the commodity form. Postone, *Time, Labor, and Social Domination*.

8. For instance: Matto Mildenberger, *Carbon Captured: How Business and Labor Control Climate Politics*, (Cambridge, MA; MIT Press, 2020); Amy Westervelt, Drilled, accessed April 23, 2021, https://www.criticalfrequency.org/drilled.

9. On life-times see the introduction, chapter 2, and Tadiar, *Remaindered Life*.

10. Laura Pulido, Ellen Kohl, and Nicole-Marie Cotton, "State Regulation and Environmental Justice: The Need for Strategy Reassessment," *Capitalism Nature Socialism* 27, no. 2 (2016): 26; also Hilda E. Kurtz, "Acknowledging the Racial State"; David N. Pellow, "Toward a Critical Environmental Justice Studies"; Laura Pulido, "Geographies of Race and Ethnicity II: Environmental Racism, Racial Capitalism and State-Sanctioned Violence," *Progress in Human Geography* 41, no. 4 (2017): 524–533. On industry participation in defining "licit" practices see: Hannah Appel, *The Licit Life of Capitalism: US Oil in Equatorial Guinea* (Durham, NC: Duke University Press Books, 2019).

11. Mark Dowie, *Losing Ground: American Environmentalism at the Close of the Twentieth Century* (MIT Press, 1996). There is a long tradition of environmentalism in the South, including advocacy by sportsmen. Loka Ashwood, *For-Profit Democracy: Why the Government Is Losing the Trust of Rural America*, (New Haven, CT: Yale University Press, 2018); Bryan, *The Price of Permanence*. Environmental justice organizing gained ground in Atlanta in the 1980s. Sellers, *Race and the Greening of Atlanta*.

12. Michael K. Heiman, "Race, Waste, and Class: New Perspectives on Environmental Justice," *Antipode* 28, no. 2 (1996): 111–121; Robert D. Bullard, ed., *Confronting Environmental Racism: Voices from the Grassroots*, First Edition (Brooklyn, NY: South End Press, 1999); Dorceta E. Taylor, "Race, Class, Gender, and American Environmentalism," Gen. Tech. Rep. (Portland, OR: US Department of Agriculture, Forest Service, Pacific Northwest Research Station, 2002).

13. Dowie, *Losing Ground*.

14. Thea Riofrancos, *Resource Radicals: From Petro-Nationalism to Post-Extractivism in Ecuador* (Durham, NC: Duke University Press Books, 2020); Kai Bosworth, *Pipeline Populism: Grassroots Environmentalism in the Twenty-First Century* (Minneapolis, MN: University of Minnesota Press, 2022).

15. On natures produced through urbanization, see Cronon, *Nature's Metropolis*; on political ecologies of electricity Andrew Needham, *Power Lines;* on the production of nature in Atlanta Tyler McCreary and Richard Milligan, "The Limits of Liberal Recognition."

16. Wright, *History of the Georgia Power Company, 1855–1956*, 82–84.

17. Manganiello, *Southern Water, Southern Power*, 43–44; on international finance see Wright, 121–128.

18. Marion M. Jackson, "Idle Slaves in the South," *The Survey*, March 1, 1924, 613, 614, 658.

19. Jackson, "Idle Slaves in the South."

20. Lennon, "Decolonizing Energy."

21. Cater, *Regenerating Dixie*, 73.

22. E. Merton Coulter, "Tallulah Falls, Georgia's Natural Wonder, From Creation To Destruction Part II," *Georgia Historical Quarterly* 47, no. 3 (1963): 267.

23. "Georgia Power Company Yearbook," 4. Powerhouses were given names inspired with by Indigenous places and history at the behest of Atkinson. See Wright, *History of the Georgia Power Company, 1855–1956*, 184.

24. Garrett, *Atlanta and Environs*, Volume II, 432.

25. "Testimony of Preston S. Arkwright Chairman of the Public Policy Committee of the National Electric Light Association before the Federal Trade Commission," 26.

26. Wright, *History of the Georgia Power Company, 1855–1956*, 113.

27. Manganiello, *Southern Water, Southern Power*, 51; see also Cobb, *The Selling of the South*.

28. Manganiello, 51.

29. Wright, *History of the Georgia Power Company, 1855–1956*, 212.

30. Wright, 332–335. In 1930, Plant Atkinson began commercial operation with both gas and coal-fired turbines, yet, required five million gallons of water every hour to cool. A drought in 1931 forced curtailments. Manganiello, *Southern Water, Southern Power*, 80–81; "Wonderful Plant Atkinson Giant of the Chattahoochee," *The Atlanta Journal*, October 16, 1930; "Plant Atkinson Now at Capacity as Water Gets Low," *The Atlanta Journal*, October 11, 1931.

31. Georgia Power Company, "Brightening People's Lives for More than 100 Years: Georgia Power Company A History in Pictures," n.d., Georgia Power Business Files, Box 1, Kenan History Center.

32. George Landry, "Special Train Transports Power," *The Macon Telegraph*, May 12, 1968; also "Opening of Coal Firm to Boost Area Economy," *Bristol Herald Courier*, December 19, 1965; Georgia Power Company, "Energy Research 1976 Annual Report," 1976, Georgia Power Business Files, Box 1, Kenan History Center; also Hall "Georgia Power Company: A Continuing Saga!!!"

33. Doyle Niemann, "Energy Conference," *Great Speckled Bird*, May 7, 1973, 5.

34. Margaret Newkirk, "Bill Would Restrict Coal Power Plants," *The Atlanta Constitution*, February 4, 2009.

35. Georgia Power Project, "Keeping Georgia Power off the Land in Monroe County," *Great Speckled Bird*, April 29, 1974, 3; Morton McInvale, "Georgia Power's Plans Bring Criticism, Praise," *The Macon News*, March 31, 1974; Morton McInvale, "Mounds Revealed on Georgia Power Site," *The Macon News*, April 21, 1974; Ken Willis, "Power Plant Threatens Indian Mounds," *The Atlanta Constitution*, April 23, 1974.

36. Interview with environmental advocate, May 25, 2023.

37. Cindy Murphy, "Georgia Power's Plant Scherer Is Fast Becoming a Reality," *The Macon News*, February 28, 1982, 15F.

38. Manganiello, *Southern Water, Southern Power*, 76.

39. "ActionLine," *The Macon News*, September 21, 1970, 1A.

40. US EPA, "Evolution of the Clean Air Act," Overviews and Factsheets, May 29, 2015, https://www.epa.gov/clean-air-act-overview/evolution-clean-air-act; Bob Hurt, "Poisoned Air Creeping Up on Atlanta," *The Atlanta Constitution*, April 21, 1969.

41. John Pennington, "The Air That Sustains Life Now Threatens It," *The Atlanta Constitution*, September 27, 1970, 8.

42. US EPA, "Evolution of the Clean Air Act."

43. Pennington, "The Air That Sustains Life Now Threatens It"; Jeff Nesmith, "Georgia Power Plant So Old Pollution Devices Won't Work," *The Atlanta Constitution*, March 16, 1970; "Dozens of Companies Violate County Rules," *The Atlanta Constitution*, April 21, 1969.

44. Nesmith, "Georgia Power Plant So Old Pollution Devices Won't Work," 13-A.

45. Charles Seabrook, "Ga. Power Using Lower-Grade Oil," *The Atlanta Journal*, March 7, 1979, sec. 12-A, https://www.newspapers.com/image/972927501/.

46. Sellers, *Race and the Greening of Atlanta*, 181–183.

47. Sellers, 183.

48. Sallye Salter, "Air Cleanup Proposal Hit by Scherer," *The Atlanta Constitution*, April 8, 1976, 12D; "Act May Force Electric Bills Up $200," *Anderson Independent*, February 5, 1978; Claudia Townsend, "33 Per Cent Rise in Electric Rates?," *The Atlanta Constitution*, April 21, 1976.

49. Townsend, "33 Per Cent Rise in Electric Rates?" 14-A.

50. "Paying Fairer Dues in the South," *The New York Times*, August 20, 1977, 20.

51. Carrie Teegardin, "Utility Spins Record Clean," *Atlanta Journal-Constitution*, June 8, 2003, A1.

52. "Group Protests Utility's Plan to Import Coal," *Alabama Journal*, May 22, 1974; quote from DeWitt Rogers, "Southern to End Coal Pact," *The Atlanta Constitution*, August 21, 1975, 9E.

53. Sam Hopkins, "Southern Co. Makes Top Coal Buy," *The Atlanta Constitution*, October 12, 1976.

54. "Utility Company's Meeting Picketed"; "Lowery Seeks End to South Africa Coal Contract," *Atlanta Voice*, May 10, 1986. Imports were eventually stopped when Congress adopted sanctions against South Africa in 1986. "Sanctions Halt Gulf Power Coal Shipments," *The Alabama Journal and Advertiser*, October 4, 1986.

55. Brian K. Obach, *Labor and the Environmental Movement: The Quest for Common Ground* (Cambridge, MA: MIT Press, 2004), 57–61. Plant Scherer was the first power plant in the Eastern US to burn "substantial amounts" of Powder River Basin coal. Falstad, "Homesick Former Montanans Buy Mall's Deck the Walls," *The Billings Gazette*, January 30, 2002.

56. Georgia Power Company's 2013 IRP and Application for Decertification of Plants Branch Units 3 and 4, Plant McManus Units 1 and 2, Plant Kraft Units 1 through 4, Plant Yates Units 1 through 5 and Plant Boulevard Units 2 and 3, Brandenburg and Hasty 81, 103 (Georgia Public Service Commission 2013)

57. Tom Kenworthy, "EPA Suspects Pollution Law Violations at Power Plants," *Washington Post*, July 16, 1999, A2.

58. Dwight Evans, "Maintenance Practices Follow the Rules," *The Atlanta Constitution*, December 20, 2001, A22; Michael Wall, "The Politics of Pollution Unmasked," *Atlanta Creative Loafing*, March 13, 2002.

59. Carrie Teegardin, "Utility's Clout Burns Bright in Washington," *The Atlanta Constitution*, February 23, 2003, A12.

60. Teegardin, A12.

61. A second round of Clean Air Act amendments in 1990 focused on ozone transport and non-attainment regions. Gloria McKinley, "Minorities Hurt the Most from Air Pollution," *The Atlanta Voice*, May 11, 1996, 1; US EPA, "Evolution of the Clean Air Act."

62. Campaign for a Prosperous Georgia et al., "Southern Fried Air: Georgia Dirty Power Plants and How to Clean Our Air Now," September 1998, ii, personal files.

63. Teegardin, "Utility Spins Record Clean."

64. Paige E. Tolbert et al., "Air Quality and Pediatric Emergency Room Visits for Asthma in Atlanta, Georgia," *American Journal of Epidemiology* 151, no. 8 (2000): 798–810; Rob McConnell et al., "Asthma in Exercising Children Exposed to Ozone: A Cohort Study," *The Lancet* 359, no. 9304 (2002): 386–391; Andrea Winquist et al., "Joint Effects of Ambient Air Pollutants on Pediatric Asthma Emergency Department Visits in Atlanta, 1998–2004:," *Epidemiology* 25, no. 5 (2014): 666–673; Jeffery M. Franklin et al., "Predictors of Emergency Department Use in Children with Persistent Asthma in Metropolitan Atlanta, Georgia," *Annals of Allergy, Asthma & Immunology* 119, no. 2 (2017): 129–136.

65. Thom Davies, "Toxic Space and Time: Slow Violence, Necropolitics, and Petrochemical Pollution," *Annals of the American Association of Geographers* 108, no. 6 (2018): 1539; McKinley, "Minorities Hurt the Most from Air Pollution."

66. Deonte and Angela Colbert quoted in Charles Seabrook, "Asthma Is a Threat That's Growing," *The Atlanta Journal-Constitution*, June 20, 1999.

67. Ann Bray quoted in Joey Ledford, "For Some, Ozone Days Are a Matter of Life and Death," *The Atlanta Journal-Constitution*, June 26, 1998.

68. Conrad Schneider, "Power to Kill: Death and Disease from Power Plants Charged with Violating the Clean Air Act," A Clear the Air Report (Boston: Clean Air Task Force, July 2001), 4, https://web.archive.org/web/20071011160828/http://www.catf.us/publications/reports/Power_to_Kill.pdf; Eric Schaeffer to Christine Whitman, "Read the Resignation Letter from Eric Schaeffer, Former Head of the U.S. EPA's Office of Regulatory," February 27, 2002, https://grist.org/article/top-epa-offical-resigns/.

69. Sellers, *Race and the Greening of Atlanta*, 311, also 295–307.

70. Pulido, Kohl, and Cotton, "State Regulation and Environmental Justice," 17.

71. Lyle V. Harris, "Justice for Our Lungs," *The Atlanta Journal-Constitution*, April 4, 2007, A14.

72. About $4 a month was added to the average residential bill to pay for scrubbers. Sarah Fay Campbell, "Scrubber Projects Planned," *The Times-Herald*, November 24, 2008, 2.

73. Doug Walker, "Hammond Is Officially Shut Down," *Marietta Daily Journal*, July 17, 2019, https://www.mdjonline.com/neighbor_newspapers/news/state/hammond-is-officially-shut-down/article_395ac3ce-543a-525a-b696-6f38c4fcec70.html.

74. Lucas Henneman et al., "Mortality Risk from United States Coal Electricity Generation," *Science* 382, no. 6673 (2023): 941–946.

75. Georgia Power, "Georgia Power's Transformational Plan for State's Energy Future Approved, Helps Ensure Company Will Continue to Meet Needs of Customers and State," PR Newswire, July 21, 2022, https://www.prnewswire.com/news-releases/georgia-powers-transformational-plan-for-states-energy-future-approved-helps-ensure-company-will-continue-to-meet-needs-of-customers-and-state-301591203.html.

76. Drew Kann, "Georgia Power's Massive Data Center Expansion Includes a Lot of Gas," *The Atlanta Journal-Constitution*, August 1, 2025, https://www.ajc.com/business/2025/08/georgia-powers-massive-data-center-expansion-includes-a-lot-of-gas/; Diana DiGangi, "Georgia Power Receives Approval to Add 1.4 GW of Natural Gas Generation," Utility Dive, April 17, 2024, https://www.utilitydive.com/news/georgia-power-approval-plant-yates-irp/713404/.

77. Dan Chapman, "Ga. Considers Own Rules for Toxic Coal Ash," *The Atlanta Constitution*, May 12, 2016, B8; Dave Williams, "Sierra Club Appealing Coal Ash Ruling to Georgia Supreme Court," *Georgia Public Broadcasting*, December 8, 2021, https://www.gpb.org/news/2021/12/08/sierra-club-appealing-coal-ash-ruling-georgia-supreme-court.

78. Abel Russ and Lisa Evans, "Georgia at a Crossroads: A Report on Coal Ash" (Environmental Integrity Project and Earthjustice, December 13, 2018), https://earthjustice.org/wp-content/uploads/ga_rpt_2018-12-12.pdf.

79. Jeff Nesmith, "Georgia Power Co. Ash Dump Charged," *The Atlanta Constitution*, September 19, 1970; Stacy Shelton, "Ash Spill Affects Rome Water Supply," *The Atlanta Constitution*, July 31, 2002; Teegardin, "Utility Spins Record Clean"; Sue Sturgis, "Coal's Ticking Timebomb: Could Disaster Strike a Coal Ash Dump Near You?," *Facing South*, January 4, 2009, https://web.archive.org/web/20160308133322/http://www.southernstudies.org/2009/01/coals-ticking-timebomb-could-disaster-strike-a-coal-ash-dump-near-you.html.

80. US EPA, "Disposal of Coal Combustion Residuals from Electric Utilities Rulemakings," Other Policies and Guidance, December 11, 2014, https://www.epa.gov/coalash/coal-ash-rule.

81. US EPA, "Disposal of Coal Combustion Residuals from Electric Utilities Rulemakings."

82. Personal communication, January 12, 2023; Liz Fabian, "Georgia Power to Spend up to $2B to Retire Coal Ash Ponds," *The Macon Telegraph*, June 21, 2016.

83. North Carolina Department of Environmental Quality, "Court Approves Consent Order to Excavate More Than 80 Million Tons of Coal Ash," February 5, 2020, https://www.deq.nc.gov/news/press-releases/2020/02/05/court-approves-consent-order-excavate-more-80-million-tons-coal-ash.

84. James Bruggers, "Congress: States Can Take Lead on Coal Ash," *The Courier-Journal*, December 16, 2016; US EPA, "Permit Programs for Coal Combustion Residual

Disposal Units," May 31, 2017, https://www.epa.gov/coalash/permit-programs-coal-combustion-residual-disposal-units; also Chapman, "Ga. Considers Own Rules for Toxic Coal Ash."

85. Max Blau, "The Coal Plant Next Door," *ProPublica*, March 22, 2021, https://www.propublica.org/article/coal-ash-georgia-power?token=Gg58888u2U5db3W3CsuKrD0LD_VQJReQ.

86. Southern Environmental Law Center, "Georgia Power Company's Applications for Solid Waste Handling Permits for Closure of Coal Ash Ponds at Plant Scherer (AP-1), Plant Wansley (AP-1), Plant Yates (AP-B and AP-3 'Ash Management Area' and R6 Landfill), Plant McDonough (AP-1, AP-3, AP-4), and Plant Hammond (AP-3)," August 5, 2019, 22, https://southernenvironment.sharefile.com/share/view/s2841595de554997a.

87. Participant observation, August 6, 2019; US EPA, "U.S. State of Georgia Coal Combustion Residuals (CCR) Permit Program," 2023, https://www.epa.gov/coalash/us-state-georgia-coal-combustion-residuals-ccr-permit-program; in 2023, the Biden administration denied a similar request from Alabama that was deemed insufficiently protective of groundwater, US EPA, "EPA Proposes to Deny Alabama's Coal Ash Permit Program and Better Protect Local Communities from Pollution," News Release, August 3, 2023, https://www.epa.gov/newsreleases/epa-proposes-deny-alabamas-coal-ash-permit-program-and-better-protect-local.

88. Catherine Morehouse, "Georgia Power Coal Ash Plan Mirrors Conditions That Led to Duke Spills: Report," *Utility Dive*, August 6, 2019, https://www.utilitydive.com/news/georgia-power-coal-ash-plan-mirrors-conditions-that-led-to-duke-spills-rep/560307/.

89. Felix Beilin, "In Georgia, Ratepayers Face a Big Bill for Coal Ash Cleanup, While a Utility Profits," *Energy News Network*, September 13, 2022, https://energynews.us/2022/09/13/in-georgia-ratepayers-face-a-big-bill-for-coal-ash-cleanup-while-a-utility-profits/.

90. Participant Observation, October 17, 2019.

91. See Appel, *The Licit Life of Capitalism*.

92. Williams, "Sierra Club Appealing Coal Ash Ruling to Georgia Supreme Court."

93. "Georgia Power Plant Hammond Ash Pond 3 Draft Permit Public Meeting" (virtual, August 11, 2021), https://www.youtube.com/watch?v=rXsryATPXY4.

94. Women living near Plant Branch also shared reproductive health concerns, see *Rising from the Ashes Chapter I: Everything Must Go Somewhere*, 2021, https://www.youtube.com/watch?v=ygXS8tprlZ0.

95. "Georgia Power Plant Hammond Ash Pond 3 Draft Permit Public Meeting," August 11, 2021.

96. Kempner, "Georgia Power's Coal Ash Plan Faces Potential U.S. EPA Pushback."

97. "Georgia Power Plant Hammond Ash Pond 3 Draft Permit Public Meeting," August 11, 2021.

98. Gilmore, "Abolition Geography and the Problem of Innocence."

99. John Sepulvado, "A Power Plant, Cancer and a Small Town's Fears," *CNN*, March 31, 2012, https://www.cnn.com/2012/03/31/us/georgia-coal-power/index.html.

100. Max Blau, "A Power Company's Quiet Land-Buying Spree Could Shield It from Coal Ash Cleanup Costs," *ProPublica*, November 24, 2020, https://www.propublica.org/article/a-power-companys-quiet-land-buying-spree-could-shield-it-from-coal-ash-cleanup-costs.

101. Blau, "The Coal Plant Next Door."

102. Alferdteen Harrison, *Black Exodus: The Great Migration from the American South* (Jackson, MS: University Press of Mississippi, 1991); Woods, *Development Arrested: The Blues and Plantation Power in the Mississippi Delta.* On the electrification of agriculture see Nye, *Electrifying America: Social Meanings of a New Technology, 1880–1940.* Also, Robin D. G. Kelley, *Freedom Dreams: The Black Radical Imagination,* (Boston: Beacon Press, 2003).

103. Marable, "Power to the People?," 13.

104. US Department of Commerce Bureau of the Census, "1980 Census of Population and Housing: Census Tracts Georgia Selected Areas," 1983, https://www2.census.gov/prod2/decennial/documents/1980/tracts-states/CensusTracts1980-Georgia.pdf; US Census Bureau, "Quickfacts: Monroe, County, Georgia," 2022, https://www.census.gov/quickfacts/fact/table/monroecountygeorgia,GA/PST045222#qf-headnote-a.

105. Ashwood, *For-Profit Democracy*, 10.

106. Ruth Wilson Gilmore, "Forgotten Places and the Seeds of Grassroots Planning," in *Engaging Contradictions: Theory, Politics and Methods of Activist Scholarship*, ed. Charles Hale (Berkeley, CA: University of California Press, 2008), 27.

107. Gilmore, 32.

108. Georgia Power Project, "Keeping Georgia Power Off the Land in Monroe County."

109. Interview with environmental organizer, May 24, 2021.

110. Harvey, building on Raymond Williams, defines militant particularism as a contradictory form of working-class organizing spurred by localized concerns that can connect to broad social movements. Militant particularism can empower organizing and also result in a defensive politics of asserting difference and uniqueness of a place, rather than seeing how injustice, like environmental pollution, is systemic. David

Harvey, *Justice, Nature, and the Geography of Difference* (Malden, MA: Blackwell Publishers, 1996), 32–45, 370–371.

111. Interview with environmental organizer, Interview May 24, 2021.

112. *Saving Juliette* (Georgia Public Broadcasting, 2021), https://www.youtube.com/watch?v=A9rK8UKOxTc.

113. Grant Blankenship, "Decades-Long Dispute over Whether Coal Ash Made Monroe County Residents Ill Is Settled out of Court," *Georgia Public Broadcasting*, December 10, 2024, https://www.gpb.org/news/2024/12/10/decades-long-dispute-over-whether-coal-ash-made-monroe-county-residents-ill-settled.

114. Interview with environmental advocate, May 24, 2021.

115. Blau, "The Coal Plant Next Door."

116. Interview with solar developer, June 4, 2019; Katherine Barnwell, "Tiny Atom Gets Slice of Research Millions," *The Atlanta Journal*, October 28, 1956, 44–45; Steve Johnson, "Solar Income Tax Credit Could Be State Bonanza, Committee Is Told," *The Atlanta Constitution*, October 19, 1979, 5C.

117. "Developers, Georgia Power to Discuss Energy Policies," *The Atlanta Constitution*, August 21, 1977, 13H; Anne Cowles, "Solar Homes to Close, Be Sold in August," *The Atlanta Constitution*, June 30, 1983; Robert Snowdon Jones, "Industrial Test of Solar Power Begins," *The Atlanta Constitution*, May 11, 1982; on electric buses see: Michael Weiss, "A Power Trip for Students," *The Atlanta Constitution*, July 18, 1996; US DOE, "Electric Buses Energize Downtown Chattanooga," August 1997, https://afdc.energy.gov/files/pdfs/chatt_cs.pdf.

118. David Anderson, Matt Kasper, and David Pomerantz, "Utilities Knew: Documenting Electric Utilities' Early Knowledge and Ongoing Deception on Climate Change from 1968–2017" (Energy and Policy Institute, July 2017), 6, https://www.energyandpolicy.org/utilities-knew-about-climate-change/; Cole Triedman et al., "American Utilities and the Climate Change Countermovement: An Industry in Flux" (Providence, RI: Brown University, Climate and Development Lab, 2019), http://www.climatedevlab.brown.edu/uploads/2/8/4/0/28401609/cdl_utilities_report_fall_2019_.pdf.

119. Pennington, "The Air That Sustains Life Now Threatens It," 8.

120. Southern Company, "2024 Annual Report"; Southern Company, "2023 Annual Report," April 16, 2024, https://s27.q4cdn.com/273397814/files/doc_financials/2023/ar/2023-annual-report.pdf.

121. Southern Company, "Form 10-K, Annual Report Pursuant to Section 13 OR 15(d) of the Securities Exchange Act of 1934 for the Fiscal Year Ended December 31, 2022," I–16.

122. Interview with energy advocate, July 30, 2019.

123. Interview with energy consultant, September 13, 2019.

124. Bill Edge, "PSC Approves Agreement to Resolve Georgia Power 2013 Integrated Resource Plan and Expands Use of Solar Energy," July 11, 2013, http://www.psc.state.ga.us/GetNewsRecordAttachment.aspx?ID=250.

125. Greg Bluestein, "Lefties and Libertarians Unite Against Common Foe: Georgia Power," *Atlanta Journal-Constitution*, August 22, 2013, https://www.ajc.com/business/lefties-and-libertarians-unite-against-common-foe-georgia-power/NYwsTm5gGmFTCPOVehoRUI/; Ethan Howard, "Georgia Power Drops Proposed Solar Fee," *Utility Dive*, November 18, 2013, https://www.utilitydive.com/news/georgia-power-drops-proposed-solar-fee/196169/; Carolyn Kormann, "Greening the Tea Party," *The New Yorker*, February 17, 2015, https://www.newyorker.com/tech/annals-of-technology/green-tea-party-solar; David Toke, *Low Carbon Politics: A Cultural Approach Focusing on Low Carbon Electricity* (Abingdon, UK: Routledge, 2018), 101–4.

126. "How Atlanta Plans to Get to 100% Green Energy by 2035," *Public Radio International*, April 15, 2019, https://www.pri.org/stories/2019-04-15/how-atlanta-plans-get-100-green-energy-2035.

127. Timinyo Owei, "The Changing Context of Energy Generation and Supply: The Case Study of Georgia" (PhD, Sheffield, UK, University of Sheffield, 2017), 265, https://core.ac.uk/download/pdf/96890576.pdf.

128. Elena Shao, "Solar Power Booms in Georgia, Where It Isn't Mandated," *Wall Street Journal*, August 22, 2021, https://www.wsj.com/articles/solar-power-booms-in-georgia-where-it-isnt-mandated-11629637200.

129. Nichola Groom and Richard Valdmanis, "Conservative Push for Solar Is Backed by Usual Green Advocates," *Reuters*, May 15, 2015, https://www.reuters.com/article/us-solar-conservatives-insight-idUSKBN0O00BT20150515; Kormann, "Greening the Tea Party."

130. Quote from "PSC Docket 42310, Docket 42311—Day 1—05/13/19: Docket 42310, Georgia Power Company's 2019 Integrated Resource Plan" (Atlanta, GA, May 13, 2019), https://www.youtube.com/watch?v=tNEdnRkDLEg&t=4807s.

131. "How Atlanta Plans to Get to 100% Green Energy by 2035."

132. Anthony Leiserowitz et al., "Politics and Global Warming, November 2019" (Yale University and George Mason University. New Haven, CT: Yale Program on Climate Change Communication, 2019), https://www.climatechangecommunication.org/wp-content/uploads/2020/01/politics-global-warming-november-2019b.pdf; Anthony Leiserowitz et al., "Politics and Global Warming, December 2020" (Yale University and George Mason University. New Haven, CT: Yale Program on Climate Change Communication, 2021), https://climatecommunication.gmu.edu/all/politics

-global-warming-december-2020/. The share of registered voters who identify as moderate Republicans and report clean energy is a high or very high priority for Congress and the president fell from 70 percent to 49 percent between 2018 and 2024. Among conservative Republicans, support fell from 54 percent to 24 percent. Anthony Leiserowitz et al., "Climate Change in the American Mind: Politics & Policy, Spring 2024" (Yale University and George Mason University. New Haven, CT: Yale Program on Climate Change Communication, 2024), https://climatecommunication.yale.edu/publications/climate-change-in-the-american-mind-politics-policy-spring-2024/.

133. Stokes, *Short Circuiting Policy*.

134. Weaver, "Georgia Power Has 5,000 Residential Solar Net Metering Contracts and It's First Come, First Served."

135. Drew Kann, "City Hall Calls Regulators' Rooftop Solar Decision 'Disappointing,'" *The Atlanta Journal-Constitution*, December 22, 2022.

136. Community solar programs are designed to allow low-income households to purchase subscriptions to offsite solar arrays that are credited to their monthly bill. Overtime, this is expected to result in customer savings. Advocates have criticized the community solar program Georgia Power established for not necessarily producing bill savings. Southern Environmental Law Center, "Post-Hearing Brief on Behalf of Georgia Interfaith Power & Light; Docket No. 44280," December 8, 2022, 17–18, https://psc.ga.gov/search/facts-document/?documentId=192355.

137. The Sierra Club, Partnership for Southern Equity, Georgia Conservation Voters, Georgia WAND, and the Southern Alliance for Clean Energy advocated for income-qualified on-bill financing for energy efficiency and the Commission approved a $7 million pilot program to provide up to $7,500 in energy efficiency measures to five hundred households in Athens and Atlanta. "Docket Nos. 42310 and 42311 in Re: Georgia Power Company's 2019 Integrated Resource Plan and Demand Side Management Plan Order Approving the Income Qualified Tariff Based Energy Efficiency Pilot with Modifications" (Georgia Public Service Commission, May 7, 2020), https://psc.ga.gov/search/facts-document/?documentId=181076. The program was launched in June 2020, however, as of 2022, only 17 installations had been completed. "State Approaches to Pay as You Save Programs" (NARUC, 2022), https://pubs.naruc.org/pub/D9C246C0-1866-DAAC-99FB-0973A12547C0. Fully subsidized programs available to individuals at or below 200 percent of the federal poverty level reached thousands of households between 2023 and 2025. Georgia Power, "Certified Demand-Side Management Programs, Third Quarter 2024 Programs Status Report; Docket No. 44161," November 15, 2024, https://psc.ga.gov/search/facts-document/?documentId=220432.

138. Carys Behnke and Taylor Shelton, "Powered by Gentrification: The Uneven Development of Residential Rooftop Solar in Atlanta, Georgia," *Energy Research & Social Science* 108 (2024): 103373.

139. Anthony Hilbert and Marion Werner, "Turn up the Heat! Contesting Energy Poverty in Buffalo, NY," *Geoforum* 74 (2016): 222–232.

140. "Direct Testimony of Francisco Valle, Andy Phillips, Jeffrey K. Smith, and Lee Evans in Support of Georgia Power Company's 2022 Integrated Resource Plan, GPSC Docket No. 44160 and Application for The Certification, Decertification and Amended Demand-Side Management Plan, GPSC Docket No. 44161" (Georgia Public Service Commission, March 11, 2022), https://psc.ga.gov/search/facts-document/?documentId=189086.

141. Georgia Power, "Georgia Power's Transformational Plan for State's Energy Future Approved, Helps Ensure Company Will Continue to Meet Needs of Customers and State."

142. Brendan Pierpont, "Clean Energy Isn't Driving Power Price Spikes" (Energy Innovation, July 2024).

143. This program builds on a customer-sited solar procurement option authorized in 2019 for which there has only been one customer. Georgia Power also applied for DOE Title 17 funding to support its Distributed Energy Resource Management System. Georgia Power, "2025 Integrated Resource Plan," January 2025, https://www.georgiapower.com/content/dam/georgia-power/pdfs/company-pdfs/2025-Integrated-Resource-Plan.pdf, 103, 109.

144. David Roberts, "Should We Put Utilities in Charge of Distributed Energy?," January 8, 2025, https://www.volts.wtf/p/should-we-put-utilities-in-charge.

145. DiGangi, "Georgia Power Receives Approval to Add 1.4 GW of Natural Gas Generation."

146. For a review, see Benjamin K. Sovacool and Michael H. Dworkin, "Energy Justice: Conceptual Insights and Practical Applications," *Applied Energy* 142 (2015): 435 444.

147. Energy Justice Network, "Environmental Justice / Environmental Racism," n.d., https://www.ejnet.org/ej/.

148. McKittrick argues knowledge systems of human life must be predicated on decolonial thinking. Katherine McKittrick, "Plantation Futures," *Small Axe: A Caribbean Journal of Criticism* 17, no. 3 42 (2013): 3.

149. Ruth Wilson Gilmore, "Forgotten Places and the Seeds of Grassroots Planning"; Tao Leigh Goffe, "Stolen Life, Stolen Time," *South Atlantic Quarterly* 121, no. 1 (2022): 109–130; on solidarity, David Roediger, "Making Solidarity Uneasy: Cautions on a Keyword from Black Lives Matter to the Past," *American Quarterly* 68, no. 2 (2016): 223–248.

Chapter 4

1. Philip Mattera et al., "High Road or Low Quality Jobs in the Green Economy?" (Washington, DC: Good Jobs First, February 3, 2009), https://web.archive.org/web/20100406201708/http://apolloalliance.org/downloads/gjfgreenjobsrpt.pdf.

2. Participant observation, September 26, 2019.

3. Stone, *Regime Politics*, 7.

4. "Union Membership and Coverage Database," 2024, https://unionstats.com/.

5. Participant observation, September 21, 2023.

6. US DOE Office of Policy, "2024 U.S. Energy & Employment Jobs Report (USEER)," 2024, https://www.energy.gov/policy/us-energy-employment-jobs-report-useer.

7. Dimitris Stevis, "Towards a Relational Environmental Labor Studies," *Journal of Industrial Relations* 64, no. 4 (2022): 608-619.

8. Cooper and Waldby, *Clinical Labor*; Geoff Mann, *Our Daily Bread: Wages, Workers, and the Political Economy of the American West* (Chapel Hill, NC: University of North Carolina Press, 2007).

9. Interview with union leader, February 13, 2023.

10. Mitchell, *Carbon Democracy*.

11. Keith Brower Brown and Sarah Holiday Nelson, "Working Sunset to Sunrise: Union Strategies in Three California Climate Transitions," *Environmental Politics* 33, no. 4 (2024): 657–77.

12. Interview with union leader, February 13, 2023; "Union Membership and Coverage Database."

13. International Labor Organization, "Guidelines for a Just Transition Towards Environmentally Sustainable Economies and Societies for All," 2015, https://www.ilo.org/wcmsp5/groups/public/---ed_emp/---emp_ent/documents/publication/wcms_432859.pdf.

14. Kenny Mullins, "Clean Energy Projects a Boon to Union Workers Making Solar Panels," *Georgia Recorder*, September 2, 2024, https://georgiarecorder.com/2024/09/02/clean-energy-projects-a-boon-to-union-workers-making-solar-panels/.

15. Liz Shuler, Yvonne Brooks, "Unions and Communities Are Building a Southern Economy for All | Opinion," *Macon Telegraph*, July 5, 2024, https://www.macon.com/opinion/article289741189.html.

16. J. Mijin Cha, "The Future of the Labor-Climate Alliance," *Dissent* 70, no. 2 (2023): 56–61; Mijin Cha, *A Just Transition for All: Workers and Communities for a Carbon-Free Future* (Cambridge, MA: MIT Press, 2024).

17. Mildenberger, *Carbon Captured*.

18. Nikki Luke, "Just Transition for All? Labor Organizing in the Energy Sector Beyond the Loss of 'Jobs Property,'" *Annals of the American Association of Geographers* 113, no. 1 (2023): 94–109. On working-class environmentalisms see Stefania Barca, "On Working-Class Environmentalism: A Historical and Transnational Overview," *Interface* 4, no. 2 (2012): 61–80; Matthew T. Huber, *Climate Change as Class War: Building Socialism on a Warming Planet* (London: Verso, 2022).

19. See Michael Ekers and Alex Loftus, "Revitalizing the Production of Nature Thesis: A Gramscian Turn?," *Progress in Human Geography* 37, no. 2 (2013): 234–252.

20. Stuart Hall, "Signification, Representation, Ideology: Althusser and the Post-Structuralist Debates," *Critical Studies in Mass Communication* 2, no. 2 (1985): 113–114.

21. Bill Fletcher and Fernando Gapasin, *Solidarity Divided: The Crisis in Organized Labor and a New Plan Toward Social Justice* (Berkeley, CA: University of California Press, 2008); also Du Bois, *Black Reconstruction in America, 1860–1880*; Thavolia Glymph, "Du Bois's *Black Reconstruction* and Slave Women's War for Freedom," *South Atlantic Quarterly* 112, no. 3 (2013): 489–505.

22. In 1845 the Georgia Legislature adopted an "Act against Black Mechanics" to restrict Black people's work as mechanics, masons, or construction. Philip S. Foner, *Organized Labor and the Black Worker, 1619–1981* (New York: International Publishers, 1982), 96–97; Leonard Hammock James, "The Policies of Organized Labor in Relation to Negro Workers in Atlanta 1869–1937" (Atlanta: Atlanta University, 1937), 3, https://dlg.usg.edu/record/auu_cau-td.

23. Tera W. Hunter, "Atlanta Washerwomen's Strike," in *Reader's Companion to U.S. Women's History* (New York : Houghton Mifflin Harcourt Publishing Company, 1998), 50.

24. Unions and their contractors developed apprenticeship programs to certify journey-level workers. These methods of training and certification were formalized through state and federal licensure. Apprenticeship is a portable credential, like a diploma. Nikki Luke et al., "Diversity in California's Clean Energy Workforce: Access to Jobs for Disadvantaged Workers in Renewable Energy Construction" (University of California, Berkeley Labor Center, 2017), http://laborcenter.berkeley.edu/diversity-in-californias-clean-energy-workforce/.

25. James, "The Policies of Organized Labor in Relation to Negro Workers in Atlanta 1869–1937," 29; Walter Drake Westmoreland, "Aspects of the History of the Negro Trade Unionist in Atlanta 1933–1942" (Masters, Atlanta, GA, Atlanta University, 1942), 6, https://citeseerx.ist.psu.edu/document?repid=rep1&type=pdf&doi=5f4d1c4357f258a24b9203309de62a2a4711c04f.

26. Westmoreland, "Aspects of the History of the Negro Trade Unionist in Atlanta 1933–1942," 5. For cases in Atlanta, see Westmoreland, "Aspects of the History of

the Negro Trade Unionist in Atlanta 1933–1942"; David R. Roediger, *Towards the Abolition of Whiteness: Essays on Race, Politics, and Working Class History* (Verso, 1994), 142; Sophia Z. Lee, *The Workplace Constitution from the New Deal to the New Right* (New York: Cambridge University Press, 2014), 49–50; on decline of biracial unionism in Alabama, see, Wilson, *America's Johannesburg*, 172–178; for broader surveys: W. E. B. Du Bois, *The Negro Artisan* (Atlanta: Atlanta University Press, 1902), 8; F. Ray Marshall, *Labor in the South* (Cambridge, MA: Harvard University Press, 1967).

27. Westmoreland, "Aspects of the History of the Negro Trade Unionist in Atlanta 1933–1942," 21–22. Westmoreland also observes technological changes that gave rise to jurisdictional disputes between unions and meant "Negro workers, more so than white workers, have not been able to keep up. . . . Because of lack of apprentices and lack of vocational guidance" (22). Apprenticeship programs in many trades were all white into the 1960s in Atlanta, Ray Marshall and Vernon M. Briggs, "Negro Participation in Apprenticeship Programs," *The Journal of Human Resources* 2, no. 1 (1967): 51–69.

28. Du Bois, *Black Reconstruction in America, 1860–1880*; James, "The Policies of Organized Labor in Relation to Negro Workers in Atlanta 1869–1937."

29. Carole E Scott and Richard D Guynn, "The Atlanta Streetcar Strikes," *Georgia Historical Quarterly* 84, no. 3 (2000): 440.

30. Kuhn, Joye, and West, *Living in Atlanta*, 72–75.

31. Kuhn, Joye, and West, 72–75.

32. On contingency in scheduling see Cooper, "Workfare, Familyfare, Godfare."

33. Scott and Guynn, "The Atlanta Streetcar Strikes."

34. O.E. Carson, *The Trolley Titans: A Mobile History of Atlanta* (Glendale, CA: Interurban Press, 1981), 75.

35. Joiner, *Let Us Reason Together: A History of Labor Relations of the Georgia Power Company*.

36. "Strikers End Long Georgia Power Dispute," *The Atlanta Journal*, January 12, 1934.

37. Joiner, *Let Us Reason Together*, 84. The company labeled the jitney service a "menace." Wright, *History of the Georgia Power Company, 1855–1956*, 153.

38. "Tracy Addresses Electrical Union," *The Atlanta Constitution*, March 2, 1940; "American Federation of Labor Leaders from Southeast Assemble for Two-Day Session," *The Atlanta Constitution*, March 3, 1940.

39. Georgia Power Company, "Annual Report 1933, Railway and Bus Transportation," 1933, MS 363, Box 2, Kenan History Center; Georgia Power Company, "Memorandum

of Agreement between Georgia Power Company and Amalgamated Association of Street, Electric Railway & Motor Coach Employees of America, Division No. 732," 1941, 17, MS 363, Box 3, Folder 3, Kenan History Center; Georgia Power Company, "Memorandum of Agreement between Georgia Power Company and Amalgamated Association of Street, Electric Railway & Motor Coach Employees of America, Division No. 732," 1943, 20, MS 363, Box 3, Folder 3, Kenan History Center.

40. IBEW business manager in 1950 quoted in IBEW Local No. 1579, "Diamond Anniversary Commemorative History," 2023, 12, https://unionhistories.com/images/uploads/books/pdf/ibew-1579-history-book-web.pdf. On skill and racialized labor hierarchies, see Evelyn Nakano Glenn, "From Servitude to Service Work: Historical Continuities in the Racial Division of Paid Reproductive Labor," *Signs: Journal of Women in Culture and Society* 18, no. 1 (1992): 1–43; Geoff Mann, *Our Daily Bread*.

41. Lane Windham, *Knocking on Labor's Door: Union Organizing in the 1970s and the Roots of a New Economic Divide* (Chapel Hill, NC: University of North Carolina Press, 2017), 20.

42. Lichtenstein, *State of the Union*, 10.

43. Windham, *Knocking on Labor's Door*, 21; Nancy Fraser, "Contradictions of Capital and Care," *New Left Review* 100, no. July–August (2016): 99–117. The Social Security Act, National Labor Relations Act, and Fair Labor Standards Act excluded agricultural and domestic workers—occupations in which Black workers were overrepresented in the South. Jill Quadagno, *The Color of Welfare: How Racism Undermined the War on Poverty* (New York: Oxford University Press, 1996); Gilmore, "Fatal Couplings of Power and Difference"; Ira Katznelson, *When Affirmative Action Was White: An Untold History of Racial Inequality in Twentieth-Century America* (New York: W. W. Norton & Company, 2006).

44. Du Bois, *Black Reconstruction in America, 1860–1880*, 597.

45. "Dixie Industrial Council Upholds 'Right to Work,'" *The Atlanta Journal*, November 21, 1945.

46. Joiner, *Let Us Reason Together*, 164.

47. Jamie Peck, "The Right to Work, and the Right at Work," *Economic Geography* 92, no. 1 (2016): 4–30.

48. A. Philip Randolph, "Right to Work Laws Called Threat to Decent Wages," in *For Jobs and Freedom: Selected Speeches and Writings of A. Philip Randolph*, ed. Andrew E. Kersten and David Lucander (Amherst, MA: University of Massachusetts Press, 1966), 108.

49. On organizing the South: Barbara S. Griffith, *The Crisis of American Labor: Operation Dixie and the Defeat of the CIO* (Philadelphia: Temple University Press, 1988).

50. On the politics of the wage: Mann, *Our Daily Bread*.

51. Today, Black workers have the highest density of union membership of any racial group. "Union Members—2023," News Release (Bureau of Labor Statistics, US Department of Labor, January 23, 2024), https://www.bls.gov/news.release/pdf/union2.pdf.

52. Monica Waugh-Benton, "Strike Fever: Labor Unrest, Civil Rights and the Left in Atlanta, 1972" (Atlanta, GA, Georgia State University, 2006), https://scholarworks.gsu.edu/history_theses/16.

53. Margaret Hurst, "Suit Charges Bias by Georgia Power," *The Atlanta Constitution*, April 13, 1968, 32.

54. Hurst, "Suit Charges Bias by Georgia Power."

55. Charles King, Affidavit, 1968, L2008–22, Box 20, Folder Deposition, International Brotherhood of Electrical Workers, Local 84 (Atlanta, GA) records, Southern Labor Archives at GSU Library.

56. Willie Moreman, interview, interview by W. M. Townsend, May 2, 1968, Tab D, L2008–22, Box 20, Folder Deposition, International Brotherhood of Electrical Workers, Local 84 (Atlanta, GA) records, Southern Labor Archives at GSU LIbrary.

57. Rufus Mitchell, Affidavit, State of Georgia, County of Cobb, 1968, L2008–22, Box 20, Folder Deposition, International Brotherhood of Electrical Workers, Local 84 (Atlanta, GA) records, Southern Labor Archives at GSU LIbrary.

58. John Swanson, Affidavit, State of Georgia, County of Fulton, 1968, 2, L2008–22, Box 20, Folder Deposition, International Brotherhood of Electrical Workers, Local 84 (Atlanta, GA) records, Southern Labor Archives at GSU LIbrary.

59. "Ga. Power Ruling Aids Black Hopefuls," *The Atlanta Constitution*, September 29, 1971, 8B; also "Judge Orders Ga. Power to Reform Racist Policies," *The Atlanta Voice*, July 10, 1971.

60. Rex Granum, "$2.1 Million Bias Judgement Hits Ga. Power," *The Atlanta Constitution*, February 1, 1974; Jon Jacobs, "Georgia Power: Paying a Little Back," *The Atlanta Voice*, February 9, 1974.

61. Tom Walker, "Will 'Atlanta Plan' Achieve Its Goals?," *The Atlanta Constitution*, June 20, 1971, 16C.

62. Joiner, *Let Us Reason Together*, 171.

63. Brenda Mooney, "Ga. Power's Seniority Plan Is Not Biased Court Rules," *The Atlanta Constitution*, May 4, 1982, 1A.

64. Walker, "Will 'Atlanta Plan' Achieve Its Goals?"

65. Ray Marshall, "Training and Entry into Union Construction" (Washington, DC: Manpower Administration, Department of Labor, 1975), 17, https://files.eric.ed.gov/fulltext/ED117379.pdf; David Sjoquist and Lakshmi Pandey, "Changing Face of

Atlanta," *Center for State and Local Finance*, 2017, https://cslf.gsu.edu/changing-face-atlanta/.

66. AFL-CIO, Civil Rights Department and Urban League, "Atlanta Plan," 1968, 30, Advancing Workers Rights in the American South: Digitizing the Records of the AFL-CIO's Civil Rights Division, Southern Labor Archive, https://digitalcollections.library.gsu.edu/digital/collection/AFLCIO/id/84328.

67. Bayor, *Race and the Shaping of Twentieth-Century Atlanta*, 49; Cynthia M. Hewitt, "African-American Concentration in Jobs: The Political Economy of Job Segregation and Contestation in Atlanta," *Urban Affairs Review* 39, no. 3 (2004): 323.

68. Hewitt, "African-American Concentration in Jobs," 336.

69. Margaret Newkirk, "Long Legal Labyrinth," *The Atlanta Constitution*, October 31, 2004.

70. Walter C. Jones, "Cochran Seeks to Expand Southern Co. Discrimination Suit," *Savannah Morning News*, April 3, 2001, 3C.

71. Matthew C. Quinn, "Power Company's Response to Bias Lawsuit Protested," *The Atlanta Constitution*, September 17, 2000, D2.

72. Quote from Russell Grantham, "Four Join Suit Alleging Bias by Southern," *Atlanta Journal-Constitution*, August 15, 2000, C2, https://www.ajc.com/business/solar-industry-waits-for-lift-from-new-law/EFfgjM8cGOIrHcSyPIqQzJ/; Matthew C. Quinn, "Bias Lawsuit Hits Georgia Power," *The Atlanta Constitution*, July 28, 2000; Matthew C. Quinn, "Ga. Power, Plaintiffs Willing to Talk," *The Atlanta Constitution*, October 13, 2001; H. D. Stewart, "Ga Power Faces Clergy," *The Atlanta Voice*, September 1, 2001; Matthew C. Quinn, "Company 'Lying' Says Cochran," *The Atlanta Journal-Constitution*, April 3, 2001.

73. Janita Poe, "Race Business," *Atlanta Journal-Constitution*, October 15, 2000, sec. Business, Q1.

74. Poe, "Race Business," Q1.

75. Poe, "Race Business," Q5.

76. Newkirk, "Long Legal Labyrinth," Q4; also Ernest Holsendolph, "Traditional 'Atlanta Way' Not Holding," *Atlanta Journal-Constitution*, October 15, 2000.

77. Newkirk, "Long Legal Labyrinth," Q8.

78. The utility audited its pay and promotion policy, instituted a diversity council, required diversity seminars for all employees, created support groups for women and workers of color, and strengthened relationships with colleges in the Atlanta University Center. Stewart, "Ga Power Faces Clergy"; Newkirk, "Long Legal Labyrinth."

79. Southern Company, "Southern Company Recognized as a Top Company for Diversity and Inclusion by DiversityInc for the Fourth Consecutive Year," May 8, 2019, http://www.southerncompany.com/newsroom/2019/may-2019/top-company-diversity-inclusion-diversityinc.html.

80. Southern Company, "Boldly Forward: Moving to Equity," 2023, 2,4, https://www.southerncompany.com/content/dam/southerncompany/sustainability/pdfs/2022-Moving-to-Equity-Report.pdf.

81. Southern Company, 8.

82. Southern Company, "Workforce Representation Disclosure: Equal Employment Opportunity Employer Information Report (EEO-1)," 2022, https://www.southerncompany.com/content/dam/southerncompany/sustainability/pdfs/SO_Workforce-Disclosure_EEO-1.pdf; see also Interstate Renewable Energy Council, "National Solar Jobs Census 2023," September 2024, https://irecusa.org/census-demographics-and-diversity/.

83. Jared Council, "Record Number of Black CEOs Will Run S&P 500 Companies," Forbes, March 1, 2023, https://www.forbes.com/sites/jaredcouncil/2023/03/01/record-number-of-black-ceos-will-run-sp-500-companies/.

84. See Jodi Melamed, *Represent and Destroy: Rationalizing Violence in the New Racial Capitalism*, Difference Incorporated (Minneapolis, MN: University of Minnesota Press, 2011); Dylan Rodríguez, *White Reconstruction: Domestic Warfare and the Logics of Genocide* (New York: Fordham University Press, 2020).

85. Following the 2020 uprisings for Black lives, Georgia Power signed onto the Metro Atlanta Chamber's "Action for Racial Equity Initiative" to "address systemic racism, inequity and immobility" in the metro region. Georgia Power, "Georgia Power Signs on to Metro Atlanta Chamber's Atl Action for Racial Equity Initiative," February 9, 2021, https://www.prnewswire.com/news-releases/georgia-power-signs-on-to-metro-atlanta-chambers-atl-action-for-racial-equity-initiative-301225428.html.

86. "A Special Supplement: Southern Power Companies," 53.

87. Bob Hall, "Georgia Power Company," *Great Speckled Bird*, July 17, 1972, 3.

88. On financial condition of Georgia Power in the 1970s, see chapter 2.

89. Hall, "Georgia Power Company," 3.

90. Hall, 3.

91. Virgil F. Neal, "Some Members Out of Work; Members Picket Square D," *IBEW Journal: Official Publication of the International Brotherhood of Electrical Workers* 71, no. 12 (December 1972): 55.

92. Virgil F. Neal, "Power Company Contracts for Union, Non-Union Work," *IBEW Journal: Official Publication of the International Brotherhood of Electrical Workers* 72, no. 3 (March 1973): 66.

93. Jim Stewart, "Rifle Fire Knocks Out Rockdale Power," *The Atlanta Constitution*, June 24, 1972; Tom Watson, "Strike! Ga. Power," *Great Speckled Bird*, June 26, 1972.

94. Hall, "Georgia Power Company," 3.

95. Hall, 3.

96. Bob Fort, "Workers Protest Request for Power Hike," *The Atlanta Constitution*, November 20, 1972, 20A; James Hightower, "New Concern for Utility: Rate Structure," *The Atlanta Journal and The Atlanta Constitution*, August 26, 1973, sec. Business News, 1E; Bill Jordan, "Power Group Rips Rate Hike," *The Atlanta Constitution*, June 13, 1973; The Georgia Power Project, "A City-Owned Power Company?," 2.

97. Interview with environmental advocate, May 25, 2023.

98. Ray Yeargain, "Scribe Highlights Proposed Nuclear Powerhouse," *IBEW Journal: Official Publication of the International Brotherhood of Electrical Workers* 73, no. 11 (November 1974): 59; Virgil F. Neal, "'Local Lines' Sing Sad Song of Nationwide Unemployment," *IBEW Journal: Official Publication of the International Brotherhood of Electrical Workers* 75, no. 1 (January 1976): 68.

99. "5th District Progress Meeting Held in Hollywood Beach, Florida," 4.

100. IBEW Local No. 1579, "Diamond Anniversary Commemorative History," 17.

101. Will Salters, "Looking Ahead," *The Electrical Worker*, May 2010, 13, https://www.ibew.org/articles/10ElectricalWorker/EW1005/LocalLines.0510.html; "After 13 Years of IBEW Work, Massive New Nuclear Reactor Goes Live," *The Electrical Worker*, October 2023, https://www.ibew.org/articles/23ElectricalWorker/EW2310/Vogtle.1023.html.

102. Interview with consumer advocate, December 3, 2020; Shelley Welton and Conor Harrison, "Lessons in Climate Derisking."

103. Interview with energy consultant, September 13, 2019

104. Jim Galloway, Greg Bluestein, and Tamar Hallerman, "The Jolt: Vogtle Nuke Project Results in a GOP Nod from Georgia AFL-CIO," *Atlanta Journal-Constitution*, September 4, 2018, https://www.ajc.com/blog/politics/the-jolt-vogtle-nuke-project-results-gop-nod-from-unions/GEI50dSlFyTy4lsY6gzpeO/.

105. Interview with environmental justice organizer, May 23, 2019; Becky Rafter, Mary Olson, and Jumana Master, "Community Impacts at the Crossroads of Nuclear and Climate Injustices in the U.S. South" (Atlanta: Georgia WAND Education Fund and Nuclear Information Resource Service, December 2017), https://gawand.org/wp-content/uploads/2017/12/GA-WAND_Climate_Nuclear_Report_Dec_2017.pdf; Paul Faeth et al., "The Water-Energy Nexus in Georgia: A Detailed Examination of Consumptive Water Use in the Power Sector" (Atlanta: Southface, Southern Environmental Law Center, Cadmus, and CNA, April 2018), https://4553qr1wvuj43kndml31ma60-wpengine.netdna-ssl.com/wp-content/uploads/2019/08/The-Water-Energy-Nexus

-in-Georgia-Study.pdf; *Quick Chats: Plant Vogtle* (Partnership for Southern Equity, 2020), https://www.facebook.com/PSEquity/videos/quick-chats-plant-vogtle/526863434677309/.

106. Interview with environmental advocate May 25, 2023.

107. Interview with union leaders, February 13, 2023.

108. Interview with consumer advocate, December 3, 2020.

109. Tom Karson, "Atlanta: Local Initiative Combines with National Support," *WorkingUSA* 10, no. 1 (March 2007): 33; Harold M Barnette and Deborah Marshall Scott, "Organizing for Progressive Power in the South," in *Igniting Justice and Progressive Power: The Partnership for Working Families Cities*, ed. David B. Reynolds and Louise Simmons (London: Routledge, 2021), 227–241. During the Olympics, the state AFL-CIO also negotiated a labor agreement for construction with targets to hire Atlanta residents. Michael Dobbins, "Race and Class in Atlanta-Style Development: Promoting Inclusion and Justice," in *Planning Atlanta* (New York: Routledge, 2014), 111–122; Stewart Acuff, "The Atlanta Labor Council: Building Power Through Mirroring the Membership," in *Central Labor Councils and the Revival of American Unionism: Organizing for Justice in Our Communities*, ed. Immanuel Ness and Stuart Eimer (Routledge, 2015), 201–220.

110. Barnette and Scott, "Organizing for Progressive Power in the South," 234.

111. Quote from Jeff Grabelsky, "'We're Getting Our Country Back': Reflections on Politics, Race, Labor and Community in the Age of Obama. A Conversation with Mark Ayers, President, Building and Construction Trades Department," *WorkingUSA* 13, no. 4 (2010): 557; also *Forum Session 1: Voices from the Field*, 2011, https://www.youtube.com/watch?v=lKmHac9y2ZE.

112. Interview with environmental advocate, May 24, 2021.

113. Interview with energy consultant, September 13, 2019; see Southern Company, "Southern Company Maintenance and Modification Agreement: Mississippi, Alabama, Georgia," August 2020, https://nabtu.org/wp-content/uploads/2017/03/SCMMA-Agreement-2020-Final.pdf.

114. Interview with labor leader, November 8, 2024.

115. Personal communication with labor leader, December 3, 2024; interview with solar industry trade group, March 10, 2025.

116. Sierra Club organizers consulted with Building Trades leaders around the 2019 IRP but did not lead any joint initiatives. Participant observation, May–July 2019.

117. Interview with energy advocate, July 30, 2019.

118. Interview with environmental advocate, May 24, 2021.

119. Lauren Kaori Gurley, "Shifting America to Solar Power Is a Grueling, Low-Paid Job," *Vice*, June 27, 2022, https://www.vice.com/en/article/z34eyx/shifting-america-to-solar-power-is-a-grueling-low-paid-job; Texas Climate Jobs Project, Cornell University School of Industrial and Labor Relations, Climate Jobs Institute, and Organized Power in Numbers, "Power and People: Working Conditions in the Texas Clean Energy Transition," 2024.

120. Solar Energy Industries Association, "Georgia Solar," Solar Energy Industries Association, 2024, https://www.seia.org/state-solar-policy/georgia-solar.

121. Interview with solar developer, August 26, 2019.

122. Interview with solar contractor, October 1, 2019.

123. Georgia Power also charged fees of up to $10,000 to review bids, which contributed to at least one solar company bankruptcy after repeated, unsuccessful bids. Interview with solar contractor, June 3, 2019.

124. Interview with solar contractor, April 29, 2019.

125. Interview with solar contractor, April 29, 2019.

126. Interview with solar contractor, October 1, 2019.

127. Interview with solar worker, December 20, 2023.

128. Interview with solar worker, August 7, 2022.

129. Gurley, "Shifting America to Solar Power Is a Grueling, Low-Paid Job"; Nik Theodore et al., "Build a Better South: Construction Working Conditions in the Southern US" (Workers Defense Project, Partnership for Working Families, 2017).

130. Interview with solar contractor, June 3, 2019.

131. Interview with solar contractor, June 3, 2019.

132. Interview with solar contractor, October 1, 2019.

133. Interview with solar contractor, April 29, 2019.

134. Interview with solar contractor, April 29, 2019.

135. Interview with solar contractor, October 1, 2019.

136. Interview with solar contractor, October 1, 2019.

137. Interview with non-union training provider, August 14, 2019.

138. Interview with solar developer, August 26, 2019.

139. Texas Climate Jobs Project et al., "Power and People: Working Conditions in the Texas Clean Energy Transition."

140. Participant observation at union open house with the US DOE, September 21, 2023; meeting with labor leaders, May 27, 2019; interview with union leader, February 13, 2023.

141. Interview with solar developer, August 26, 2019.

142. Interview with solar developer, August 15, 2019

143. Interview with solar developer, September 23, 2019.

144. Interviews with former city official, June 2, 2019; solar developer, September 23, 2019.

145. Personal communication with solar contractor, June 2, 2025.

146. Interview with labor leader February 13, 2023.

147. Emily Pontecorvo, "The IRA's Labor Provisions Look Like They're Working," Heatmap, August 15, 2024, https://heatmap.news/economy/ira-anniversary-labor; calculation of private construction union membership over time based on data available since 1986 from Barry T. Hirsch, David A. Macpherson, and William E. Even, "Union Membership and Coverage Database," 2024, https://www.unionstats.com/.

148. Interview with apprenticeship coordinator, October 2, 2024.

149. Interviews with apprenticeship coordinators, October 2, 2024; October 24, 2024.

150. Interview with apprenticeship coordinator, October 24, 2024.

151. "Gov. Lee Joins Coalition of Governors in Opposing UAW's Unionization Campaign," TN Office of the Governor, April 16, 2024, https://www.tn.gov/governor/news/2024/4/16/gov--lee-joins-coalition-of-governors-in-opposing-uaw-s-unionization-campaign.html. In a 2021 speech to Georgia's congressional delegation, Kemp framed equity and labor standards in draft federal climate policy as an existential threat to Georgia's EV industry, cautioning "you have a choice: jobs for hardworking Georgians or jobs for labor unions." Brian P. Kemp to Members of Georgia's Congressional Delegation, October 8, 2021, letter, personal files. In his 2024 state of the state address, Kemp pilloried President Biden's support of organized labor as a "war on opportunity." Delaney Tarr, "Governor Brian Kemp to Tackle Unions, Transportation and Education in 2024," *SaportaReport*, January 11, 2024, http://saportareport.com/governor-brian-kemp-to-tackle-unions-transportation-and-education-in-2024/sections/reports/delaneytarr/.

152. Michelle Baruchman, "Kemp Signs Bill Discouraging Unionization in Georgia," *The Atlanta Journal-Constitution*, April 24, 2024.

153. Interview with workforce development organization, June 25, 2019.

154. Interview with labor leader, February 13, 2023.

155. Jonathan Weisman, "Labor Groups Target Hyundai—and Biden—Over Transition to Electric," *The New York Times*, August 27, 2023, https://www.nytimes.com/2023/08/27/us/politics/georgia-hyundai-labor-biden.html.

156. *PSC Docket 56002 Georgia Power IRP & 56003*, 2025, https://www.youtube.com/watch?v=FGngHPoZxEY.

157. Interview with labor leader, February 13, 2023; Participant observation, September 21, 2023.

158. Interview with community-labor organizer, August 21, 2024.

159. Du Bois, *Black Reconstruction*, 630–631.

Chapter 5

1. Anoop Mirpuri, Keith P. Feldman, and Georgia M. Roberts, "Antiracism and Environmental Justice in an Age of Neoliberalism: An Interview with Van Jones," *Antipode* 41, no. 3 (2009): 408–409.

2. Van Jones, *The Green Collar Economy: How One Solution Can Fix Our Two Biggest Problems* (New York: HarperOne, 2008), 17.

3. Sarah Knuth, "Whatever Happened to Green Collar Jobs? Populism and Clean Energy Transition," *Annals of the American Association of Geographers* 109, no. 2 (2019): 634–643.

4. Sarah White, Joel Rogers, and Laura Dresser, "Greener Reality: Jobs, Skills, and Equity in a Cleaner U.S. Economy," 2012, 20, https://www.cows.org/greener-reality-jobs-skills-and-equity-in-a-cleaner-us-economy_1. A court settlement with electricity providers that rigged California's newly deregulated electricity market in 2001 funded the first Green Jobs Corps. This seed money supported the Ella Baker Center, the IBEW, and the Apollo Alliance to partner on pre-apprenticeship programs for people of color under-represented in the Building Trades. Kim, Kirsch, and Reyes, "Making Green Work: Best Practices in Green-Collar Job Training."

5. Interviews with Wisconsin workforce training program April 14, 2016; Illinois environmental justice organizations, May 3, 2016; May 5, 2016; Minnesota clean energy contractor, June 22, 2016; Minnesota community development corporation, June 21, 2016; interview with California workforce training provider, February 1, 2018.

6. Personal communication with labor consultant, May 17, 2019.

7. Mirpuri, Feldman, and Roberts, "Antiracism and Environmental Justice in an Age of Neoliberalism," 404–5; see also Jones, *The Green Collar Economy*.

8. United Nations Department of Economic and Social Affairs, "A Global Green New Deal for Climate, Energy, and Development," December 2009, https://sustainable

development.un.org/content/documents/cc_global_green_new_deal.pdf; see Stuart Hall, Doreen Massey, and Michael Rustin, eds., *After Neoliberalism? The Kilburn Manifesto* (London: Lawrence & Wishart, 2015), 130.

9. Mattera et al., "High Road or Low Quality Jobs in the Green Economy?," 5.

10. Apollo Alliance, "The New Apollo Program: Clean Energy Good Jobs," September 2008, 16, https://web.archive.org/web/20090306083041/http://www.apolloalliance.org/downloads/fullreportfinal.pdf.

11. Personal communication with former Apollo Alliance staff person, January 13, 2025.

12. David R. Obey, "Text—H.R.1—111th Congress (2009–2010): American Recovery and Reinvestment Act of 2009" (2009), https://www.congress.gov/bill/111th-congress/house-bill/1/text; Jeannie Choi, "The Green Industrial Revolution," *Sojourners Magazine*, May 1, 2009; US Department of Labor, "American Recovery and Reinvestment Act of 2009: Green Capacity Building Grants," 2009, https://www.doleta.gov/pdf/GreenJobs.pdf. On Green Jobs Act, see Thomas Friedman, "Win over Blacks with Green Jobs," *The Atlanta Constitution*, October 21, 2007; "NAACP Support for Present and Future Green Jobs Appropriations and Policies," NAACP, January 1, 2008, https://naacp.org/resources/naacp-support-present-and-future-green-jobs-appropriations-and-policies.

13. On union green jobs campaigns, see Grabelsky, "'We're Getting Our Country Back'"; Fine, "When the Rubber Hits the High Road."

14. US Bureau of Labor Statistics, "Measuring Green Jobs," Bureau of Labor Statistics, 2013, https://www.bls.gov/green/home.htm.

15. Interview with workforce development funder, May 16, 2019.

16. Interview with solar industry trade group, March 10, 2025.

17. Alyssa Battistoni, "Living, Not Just Surviving," *Jacobin*, 2017.

18. City of Atlanta Mayor's Office of Resilience, "Clean Energy Atlanta: A Vision for a 100% Clean Energy Future," 52; Raisa Habersham, "City OKs Facilities Clean Energy Plan," *The Atlanta Constitution*, March 5, 2019.

19. Jamie Peck, *Workfare States* (New York: The Guilford Press, 2001).

20. Jamie Peck, "Workfare: A Geopolitical Etymology," *Environment and Planning D: Society and Space* 16, no. 2 (1998): 133–161; Gerstle, *The Rise and Fall of the Neoliberal Order*.

21. Sam Strong, "The Work of Looking for Work: Surviving Without a Wage in Austerity Britain," in *Beyond the Wage: Ordinary Work in Diverse Economies*, ed. William Monteith, Dora-Olivia Vicol, and Philippa Williams (Bristol, UK: Policy Press, 2021).

22. R. Werum, "Matching Youth and Jobs? Gender Dynamics in New Deal Job Training Programs," *Social Forces* 81, no. 2 (2002): 473–503.

23. Rutheiser, *Imagineering Atlanta*, 21; Driskell, *Schooling Jim Crow*.

24. Karen Ferguson, *Black Politics in New Deal Atlanta* (Chapel Hill, NC: The University of North Carolina Press, 2002), 83, 90.

25. Judith Stein, "'Of Mr. Booker T. Washington and Others': The Political Economy of Racism in the United States," *Science & Society* 38, no. 4 (1974/1975): 442.

26. Stein, 443.

27. Nancy Rose, *Workfare or Fair Work: Women, Welfare, and Government Work Programs* (New Brunswick, NJ: Rutgers University Press, 1995).

28. Booker T. Washington, "Booker T. Washington Delivers the 1895 Atlanta Compromise Speech," in *The Booker T. Washington Papers, Volume 3*, ed. Louis R. Harlan (Urbana, IL: University of Illinois Press, 1895), 583–587.

29. "A Plea for His Race," *The Atlanta Constitution*, September 19, 1895.

30. W.E.B. Du Bois, *The Souls of Black Folk* (Millennium Publications, 1903), 24, 27; on industrial education programs see Booker T. Washington, "Industrial Training for the Negro" (Tuskegee Institute Steam Print, 1904), https://babel.hathitrust.org/cgi/pt?id=uc1.31175035177552;view=1up;seq=16.

31. Du Bois, *The Souls of Black Folk*, 30.

32. Rose, *Workfare or Fair Work*.

33. Quoted in Ferguson, *Black Politics in New Deal Atlanta*, 75.

34. Ferguson, *Black Politics in New Deal Atlanta*.

35. Ferguson, 104.

36. Ferguson. Black union members organized for jobs on federal housing projects, as did the Urban League, which also helped secure hiring for Black workers in defense contracting during World War II. Westmoreland, "Aspects of the History of the Negro Trade Unionist in Atlanta 1933–1942"; Merl E. Reed, "Bell Aircraft Comes South: The Struggle by Atlanta Blacks for Jobs During World War II," in *Labor in the Modern South*, ed. Glenn T. Eskew (Athens, GA: University of Georgia Press, 2001), 102.

37. Westmoreland.

38. The GI Bill paid for veterans to attend high school, vocational training, or college through federal grants disbursed to local and state officers. Black veterans in the South often received reduced benefits. Charles G. Bolte and Louis Harris, *Our Negro Veterans* (New York: Public Affairs Committee, Inc., 1947).

39. Garth L. Mangum, *MDTA: Foundation of Federal Manpower Policy* (Baltimore, MD: The John Hopkins Press, 1968); Gladys Roth Kremen, "MDTA: The Origins of the Manpower Development and Training Act of 1962," US Department of Labor, 1974, https://www.dol.gov/general/aboutdol/history/mono-mdtatext.

40. Bayor, *Race and the Shaping of Twentieth-Century Atlanta*, 119.

41. Rose, *Workfare or Fair Work*, 76–83. Training programs included curricula from basic literacy to advanced manufacturing skills but were not a comprehensive labor policy and had "little capacity to alter patterns of private activity, selective interventions in employment had little to do with American economic policy; they became viewed as welfare policies that offered minimal services to the poorest citizens." Margaret Weir, "The Federal Government and Unemployment: The Frustration of Policy Innovation from the New Deal to the Great Society," in *The Politics of Social Policy in the United States*, ed. Margaret Weir, Ann Shola Orloff, and Theda Skocpol (Princeton, NJ: Princeton University Press, 1988), 150. A 1968 analysis of the Manpower Act demonstrated institutional training struggled to correct the "'last in-first out' problems" of "various disadvantaged groups." Mangum, *MDTA: Foundation of Federal Manpower Policy*, 95–96, 98.

42. Ada Biehl, "Some Pluses, Minuses of Computer Training," *The Atlanta Constitution*, July 13, 1970, 3B.

43. Mary Kay Murphy, "Life's Work: Strong Backs Cannot Find Jobs," *The Atlanta Constitution*, April 5, 1968, 31.

44. Hill, "Labor Unions and the Negro"; Judith Stein, "Affirmative Action and the Conservative Agenda: President Richard M. Nixon's Philadelphia Plan of 1969," in *Labor in the Modern South*, ed. Glenn T. Eskew (Athens, GA: University of Georgia Press, 2001), 102–34; Thomas J. Sugrue, "Affirmative Action from Below: Civil Rights, the Building Trades, and the Politics of Racial Equality in the Urban North, 1945–1969," *Journal of American History* 91, no. 1 (2004): 145–173; Adolph Reed Jr., "Judith Stein and the Historical Materialist Study of American Political History," *Nonsite.Org*, May 10, 2019, https://nonsite.org/article/judith-stein-and-the-historical-materialist-study-of-american-political-history.

45. See Peck, *Workfare States*; Rose, *Workfare or Fair Work*.

46. Interview with former congressional staffer, July 10, 2019.

47. "Jobs and Prices in the West Coast Region" (Washington, DC: US Government Printing Office, January 12, 1976), 31.

48. Patrick Heffernan, "Jobs for the Environment-The Coming Green Collar Revolution" (Joint Economic Committee, 94th Congress of the United States, 1976), 134.

49. Interview with researcher, June 28, 2019. Notably, the California Conservation Corps was also established in 1976 from the California Ecology Corps created by

Governor Reagan as a work program for conscientious objectors to the Vietnam War. Clean energy was also identified as an employment opportunity. Fred Banfman and Ed Smeloff, *Jobs from the Sun*, 1979.

50. "Jobs and Prices in the West Coast Region."

51. "Jobs and Prices in Atlanta" (Washington, DC: US Government Printing Office, December 6, 1975), 8.

52. "Jobs and Prices in Atlanta," 103.

53. Interview with researcher, June 28, 2019; also Derek Shearer, "'Green' Economics: Planning with a Political Face," *The Nation*, 1983.

54. Interview with researcher, June 28, 2019.

55. Jefferson R. Cowie, *Stayin' Alive: The 1970s and the Last Days of the Working Class* (New York: The New Press, 2010), 262, also 270–288.

56. Eva Bertram, *The Workfare State: Public Assistance Politics from the New Deal to the New Democrats* (Philadelphia: University of Pennsylvania Press, 2015).

57. CETA employed nearly 750,000 adults and created a million summer jobs for teenagers in the 1970s; by 1982, the summer employment program was halved, adult employment eliminated, and the policy framework replaced by local pilots and federal workfare initiatives. Maureen A. McLaughlin, "Effects of Eliminating Public Service Employment," Staff Working Paper (Washington, DC: Congressional Budget Office US Congress, 1981); Gordon Lafer, *The Job Training Charade* (Ithaca, NY: Cornell University Press, 2002).

58. Lafer, *The Job Training Charade*, 21.

59. Lafer, 5.

60. This is a pseudonym.

61. Interview with workforce training administrator, July 24, 2019.

62. Interview with workforce training administrator, July 24, 2019.

63. Interview with workforce training provider, September 13, 2019.

64. Harry Braverman, *Labor and Monopoly Capital: The Degradation of Work in the Twentieth Century*, (New York: Monthly Review Press, 1998). Braverman argues increasing educational requirements rarely promote intellectual curiosity or education, or reflect the changing competencies needed for work, but rather manage a surplus labor reserve.

65. Gramsci, *Selections from the Prison Notebooks*, 350; Paul E. Willis, *Learning to Labour: How Working Class Kids Get Working Class Jobs* (Farnborough, UK: Saxon House, 1977).

66. Peck, "Workfare: A Geopolitical Etymology"; Bertram, *The Workfare State: Public Assistance Politics from the New Deal to the New Democrats.*

67. Cooper, *Family Values*, 109.

68. Isabella Bakker and Stephen Gill, eds., *Power, Production and Social Reproduction: Human In/Security in the Global Political Economy* (New York: Palgrave Macmillan, 2003); Katz, "Bad Elements."

69. Michael K. Brown, "Ghettoes, Fiscal Federalism, and Welfare Reform," in *Race and the Politics of Welfare Reform*, eds. Sanford F. Schram, Joe Soss, and Richard C. Fording (Ann Arbor, MI: University of Michigan Press, 2003), 65–66.

70. Interview with technical college administrator, September 23, 2019.

71. Interview with workforce training administrator, July 24, 2019.

72. On the collapse of social welfare in Atlanta public housing, see Rodriguez, *Diverging Space for Deviants.*

73. Tadiar, *Remaindered Life*; Mullings, "Caliban, Social Reproduction and Our Future Yet to Come"; on organized abandonment, see David Harvey, *The Limits to Capital* (London: Verso, 2006), 397; Gilmore, "Forgotten Places and the Seeds of Grassroots Planning"; Rodriguez.

74. Brown, *Undoing the Demos,* 33.

75. Peck, *Workfare States*; Cooper, *Family Values.*

76. Interview with workforce development funder, May 16, 2019.

77. See Dana-Ain Davis, "Manufacturing Mammies: The Burdens of Service Work and Welfare Reform Among Battered Black Women," *Anthropologica* 46, no. 2 (2004): 282.

78. Opportunity youth was intended to be a "less-pejorative" way to categorize underemployed young people who might participate in criminalized activities. Alexander Camardelle, "Illusions of Opportunity? A Critical Discourse Analysis of Atlanta's Opportunity Youth Initiative" (PhD, Atlanta, GA: Georgia State University, 2021), 2, 124.

79. The term *green collar jobs* was used in the Pacific Northwest. Alan T. Durning, *Green Collar Jobs: Working in the New Northwest* (Seattle: Sightline Institute, 1999); for a national genealogy, see Knuth, "Whatever Happened to Green Collar Jobs?"

80. "'Millions of Green Jobs to Be Created,'" *Financial Times*, July 14, 1992; "Presidential Earth Day Address" (C-SPAN, April 21, 1993), https://www.c-span.org/video/?39911-1/presidential-earth-day-address&start=659.

81. William J. Clinton, "Executive Order 12898 of February 11, 1994, Federal Actions To Address Environmental Justice in Minority Populations and Low-Income Populations," *Federal Register* 59, no. 32 (February 16, 1994): 7629–7633.

82. Bullard, *Dumping in Dixie: Race, Class, and Environmental Quality*, 10.

83. Marshall Latimore, "City of Atlanta Meets 'Better Buildings Challenge' Goal 2 Years Ahead of Schedule," *The Atlanta Voice*, November 1, 2019.

84. *Forum Session 1: Voices from the Field.*

85. Interview with energy advocate, July 30, 2019.

86. Interview with energy advocate, July 30, 2019.

87. Leon Stafford, "Reed Touts Center as a 'Green' Job Creator," *The Atlanta Constitution*, October 30, 2010; interviews with energy justice advocate April 15, 2019; city employee, March 20, 2019.

88. YouthBuild is a US Department of Labor-funded program for young people to complete their GED and start vocational training while learning construction, or another industry.

89. Interview with workforce training provider, September 13, 2019.

90. Interview with workforce training provider, September 13, 2019; one continuing program was the Department of Watershed Management's five-week Care and Conserve plumbers' helper program that funds nonprofits to install plumbing repairs and water conservation measures. Interviews with city employees, May 21, 2019.

91. Interview with energy justice advocate, April 15, 2019.

92. Interviews with city employees, March 20, 2019; May 21, 2019; energy justice advocate, April 15, 2019.

93. Interview with workforce training provider, September 13, 2019.

94. Interview with workforce training provider, June 25, 2019.

95. Interview with community development corporation, June 21, 2016.

96. Interview with federal administrator, August 6, 2019.

97. Dustin Mulvaney, *Solar Power: Innovation, Sustainability, and Environmental Justice* (Oakland, CA: University of California Press, 2019).

98. The project was a spin off from Georgia Tech's University Center of Excellence in Photovoltaics, "Project Profile: Suniva," Quick Start Technical College System of Georgia, 2019, https://www.georgiaquickstart.org/suniva.

99. Interview with technical college administrator, September 23, 2019.

100. Alison F. Takemura, "This Georgia Program Is Training a Huge Cleantech Manufacturing Workforce," *Canary Media*, June 9, 2023, https://www.canarymedia.com/articles/clean-energy-jobs/this-georgia-program-is-training-a-huge-cleantech-manufacturing-workforce.

101. Tedd Brown, "Helping Hands," *The Columbus Ledger*, January 2, 1972; H. M. Cauley, "Turning Our High-Tech Workers: State Getting Busy with Training," *The Atlanta Constitution*, April 5, 1999.

102. Interview with technical college administrator, September 23, 2019.

103. Interview with technical college administrator, September 23, 2019.

104. "SK Battery America Commerce," Georgia Department of Labor, 2020, https://dol.georgia.gov/events/2020-02-29/sk-battery-america-commerce; Jeff Gill, "With 2,000 Jobs Expected by 2025, SK Battery Holding Feb. 29 Job Fair," *The Times*, February 14, 2020, https://www.gainesvilletimes.com/news/business/2000-jobs-fill-sk-battery-holding-feb-29-job-fair/; Parrish Walton, "500 Jobs, $150 Million In Investment Coming To Whitfield County," *Georgia Public Broadcasting*, June 4, 2018, https://www.gpb.org/blogs/georgia-at-work/2018/06/04/500-jobs-150-million-investment-coming-whitfield-county.

105. Interview with workforce training provider, September 13, 2019.

106. Interview with workforce training provider, September 13, 2019.

107. Interview with labor-community organizer, March 26, 2019.

108. "Electric Transportation Ecosystem" (Georgia Power Economic Development, 2024), https://www.selectgeorgia.com/discover-georgia/industries/EV-ecosystem-in-georgia/.

109. Timothy J. Minchin, "When Kia Came to Georgia: Southern Transplants and the Growth of America's 'Other' Automakers," *Journal of Southern History* 83, no. 4 (2017): 889–930.

110. A retired union leader commented that Governor Perdue failed to support to Ford and GM plants in Atlanta that closed in the same period in his enthusiasm to rid the state of the United Auto Workers. Interview November 8, 2024.

111. Ron Starner, "Georgia: Georgia Supercharges Growth Through Worker Training," *Site Selection*, 2023, https://siteselection.com/SpecialReports/workforce/2023/georgia-supercharges-growth-through-worker-training.cfm.

112. Dan Chapman, "West Point, Region Eager to Reap Kia Plant's Yield," *The Atlanta Constitution*, November 8, 2009, A14.

113. David Shepardson, "Kia to Invest $200 Million in US Plant, Build EV9 SUV in 2024," *Reuters*, July 13, 2023, https://www.reuters.com/business/autos-transportation/kia-invest-200-million-georgia-plant-build-ev9-suv-2024-2023-07-12/.

114. Starner, "Georgia."

115. Starner.

116. Starner.

117. Takemura, "This Georgia Program Is Training a Huge Cleantech Manufacturing Workforce."

118. Personal communication with Georgia-based energy company executive, September 27, 2023.

119. Georgia USA, "Governor Kemp Welcomes Rivian for Single-Largest Economic Development Project in State History," December 16, 2021, https://georgia.org/press-release/governor-kemp-welcomes-rivian-single-largest-economic-development-project-state; Georgia USA, "Hyundai Motor Group to Invest $5.54 Billion in Georgia at First Fully Dedicated Electric Vehicle and Battery Manufacturing Facility," May 20, 2022, https://georgia.org/press-release/hyundai-motor-group-invest-554-billion-georgia-first-fully-dedicated-electric-vehicle.

120. Georgia Department of Economic Development, "Summary: Economic Development Agreement with Hyundai Motor Group," 2022, https://georgia.org/sites/default/files/2022-07/hyundai_motor_group_economic_development_agreement_-_executive_summary.pdf; "Economic Development Agreement Rivian Horizon, LLC and the State of Georgia and the Georgia Department of Economic Development; and Joint Development Authority of Jasper County, Morgan County, Newton County and Walton County," May 2, 2022, 34, C-1–C-4, https://www.stantonsprings.com/uploads/7/5/0/2/75024267/jda-rivian_--_economic_development_agreement-_fully_executed_5.2.22.pdf.

121. Rivian, "Rivian Launches Apprenticeship Program to Support Georgia Plant," October 27, 2023, https://rivian.com/newsroom/article/rivian-launches-apprenticeship-program-to-support-georgia-plant.

122. "Economic Development Agreement Rivian Horizon, LLC and the State of Georgia and the Georgia Department of Economic Development; and Joint Development Authority of Jasper County, Morgan County, Newton County and Walton County," 2.

123. Georgia is among ten states that have not taken the Medicaid expansion. It did establish the nation's only work requirement for Medicaid. MaryBeth Musumeci, Elizabeth Leiser, and Megan Douglas, "Few Georgians Are Enrolled in the State's Medicaid Work Requirement Program," The Commonwealth Fund, September 11, 2024.

124. John Ruch, "Rivian Project Hit with State Violation, Local Lawsuit in Dirt Pollution Controversy," *SaportaReport*, October 25, 2022, http://saportareport.com/rivian-project-hit-with-state-violation-local-lawsuit-in-dirt-pollution-controversy/sections/reports/johnruch/.

125. Tadiar, *Remaindered Life*, 114.

126. Alexander Burns, "Brian Kemp and the Electric Car: A Love Story," *Politico*, January 13, 2023, https://www.politico.com/news/magazine/2023/01/13/brian-kemp-electric-car-georgia-00077579.

127. Interview with workforce training provider, September 13, 2019.

128. Georgia Power, "Lineworker Entry Program," 2025, https://www.georgiapower.com/about/company/careers/lineworker-entry.html.

129. On the concomitant transformation of higher education that mirrors changes in training programs, Cooper, *Family Values*; Hannah Appel, "The Debt Collective: $600 Million in Debt Discharge," Challenge Inequality, accessed June 24, 2023, https://challengeinequality.luskin.ucla.edu/the-debt-collective-600-million-in-debt-discharge/.

Conclusion

1. Participant observation, March 14, 2019.

2. Jelani Cobb, "William Barber Takes on Poverty and Race in the Age of Trump," *The New Yorker*, May 7, 2018, https://www.newyorker.com/magazine/2018/05/14/william-barber-takes-on-poverty-and-race-in-the-age-of-trump.

3. *Ella Baker: Making the Struggle Every Day*, 2017, https://www.youtube.com/watch?v=t96fnyLMihA.

4. "Interview with Ella Baker," *The Urban Review* 4, no. 3 (1970): 23.

5. Nixon, "The Changing Political Philosophy of the South"; Chandra Childers, "Southern Economic Policies Undermine Job Quality for Auto Workers."

6. Andreas Folkers, "Future Theft: Fossil Capitalization, Ecological Dispossession, and Climate Reparations," *Theory, Culture & Society*, 2025.

7. Gilmore, "Abolition Geography and the Problem of Innocence," 227.

8. "Georgia Power Project," *Great Speckled Bird*, 15.

9. See Chan and Klass, "Regulating for Energy Justice."

10. Fairchild and Weinrub, *Energy Democracy*, 18.

11. On organizing and the state, see Ruth Wilson Gilmore, Ruth Wilson Gilmore on Covid-19, Decarceration, and Abolition, interview by Naomi Murakawa, Haymarket Books, April 17, 2020, https://www.haymarketbooks.org/blogs/128-ruth-wilson-gilmore-on-covid-19-decarceration-and-abolition.

12. *PSC Dockets #56002 & #56003 Georgia Power Company 2025 Plans & Applications*.

13. Noel Castree and Brett Christophers, "Banking Spatially on the Future: Capital Switching, Infrastructure, and the Ecological Fix," *Annals of the Association of*

American Geographers 105, no. 2 (2015): 378–386; Patrick Robbins et al., "Who Owns Power in the Energy Transition?" *Climate and Community Project*, July 2024.

14. Cater, *Regenerating Dixie*.

15. Farrell, "Boulder Voters Say (Again): 'We'll Lead Movement to Energy Democracy'"; John Farrell, "Minneapolis' Once-a-Decade Chance to Fight for Energy and the Climate," *Star Tribune*, June 15, 2025, sec. Opinion Exchange.

16. Christophers, *The Price Is Wrong*, 376–78.

17. Judy Lin, "What Happens If California Takes over PG&E?," *CalMatters*, February 5, 2020, http://calmatters.org/politics/2020/02/what-happens-if-california-takes-over-pge/.

18. Russ Mitchell, "Is California Government Considering Oil Refinery Takeovers? Yes, It Is," *Los Angeles Times*, February 16, 2025, https://www.latimes.com/environment/story/2025-02-16/is-california-government-considering-oil-refinery-takeovers-yes-it-is.

19. Sandeep Vaheesan, *Democracy in Power: A History of Electrification in the United States* (Chicago: University of Chicago Press, 2024). The Georgia Power Project also questioned the organization of the utility, arguing that customers and workers should be on the board "to insure that our needs become predominant over the interests of the profit-seekers." "Georgia Power Project," *Great Speckled Bird*, September 11, 1972, 15.

20. Mariame Kaba, *We Do This 'Til We Free Us: Abolitionist Organizing and Transforming Justice*, ed. Tamara K. Nopper (Chicago: Haymarket Books, 2021), 25.

21. Interviews with energy democracy organizers, 2016; reproductive justice organizer, October 21, 2021; gender and economic justice organization president, August 5, 2022; environmental advocate, May 25, 2023.

22. Lowe, *The Intimacies of Four Continents*, 21; see also Deborah Cowen, "Following the Infrastructures of Empire: Notes on Cities, Settler Colonialism, and Method," *Urban Geography*, (2019), 1–18.

23. Gilmore argues that "nonfatal power-difference couplings" require attention to mutuality to work toward the abolition of systems that reproduce vulnerability to premature death. Gilmore, "Fatal Couplings of Power and Difference," 16.

24. Bobby Seale, *Seize The Time: The Story of the Black Panther Party* (Baltimore, MD: Black Classic Press, 1991), 233.

Index

Abernathy, Juanita, 27–28
Advertising, 63–67, 71
Affirmative action, 123–124, 127
Affordability, of energy, 1, 8, 15, 55, 61, 69, 77–80, 89, 111, 167, 171. *See also* assistance; racialized uneven development; rates
  energy burden, 1, 32, 43, 55, 77, 79, 88, 116, 129–130, 171
  prepay meters, 81–82
AFL-CIO, Georgia, 128–129
Air conditioning, 67, 82, 202n38
Amalgamated Association of Street and Electric Railway Employees, 119, 120
American Recovery and Reinvestment Act (ARRA) (2009), 52, 140
American Rescue Plan Act (2021), 53
Angel, James, 24
Apartheid, South Africa, 98
Apollo Alliance, 139, 140
Apprenticeship, 114–115, 118, 129–130, 134–135, 141, 146, 159, 162, 223n24, 224n27, 233n4
Arkwright, Preston, 11, 12–13, 43, 63, 93, 119, 121
Ashwood, Loka, 104–105
Assistance, energy, 62, 78–79. *See also* affordability, of energy; social reproduction
Asthma, 90, 99–100, 108, 169
Atkinson, H. M., 36–37, 143
Atlanta, City of, 34–41, 49, 73–74, 83, 153
  100 percent clean energy resolution (Clean Energy Atlanta), 8–9, 31–33, 40, 52, 54, 55, 88, 141, 155, 170, 171
  Board of Aldermen/City Council, 36, 39, 41, 43, 52, 73, 98
  Equal Employment Opportunity, 123
  Office of Sustainability and Resilience, 31–32, 51–52, 141
  Public Safety Training Center ("Cop City"), 54
  sustainability plans, 50–55, 154
Atlanta Constitution (newspaper), 124, 144, 146
Atlanta Cotton States and International Exposition, 143–144
Atlanta Journal-Constitution (newspaper), 76, 79, 99, 158
Atlanta Railway and Power Company, 36–37
Atlanta Regional Commission, 49
Atlanta Race Massacre, 39–41
Atlanta University Center, 34, 118, 143
Atlanta Way, the, 33, 44, 122–124
Austerity, 50, 52, 147, 200n9

Baker, Ella, 25–26, 139, 166–167
Barber, William, 165–167

Barr, Bob, 99
Beltline, Atlanta, 51, 55
Better Buildings Challenge, 53, 130, 153–154
Bhattacharyya, Gargi, 21
Biden administration, 59–61, 113–115, 136, 159. *See also* Bipartisan Infrastructure Law; Inflation Reduction Act
Bipartisan Infrastructure Law (BIL) (2021), 4, 59, 113, 115
Black Lives Matter movement, 55, 106, 228n85
Black Reconstruction (Du Bois), 34–35, 120
Blue Bird electric bus factory, 115
BlueGreen Alliance, 139
Bonbright, James, 12
Boosterism, 28, 33–34, 50, 54
Bottoms, Keisha Lance, 53–54
Bowers, Paul, 81
Branch, Harlee, Jr., 97
Brush Electric Company, 35
Bullard, Robert, 152–153
Burgess, David, 56
Busbee, George, 97
Bush, George W. administration, 99, 100

Campaign for a Prosperous Georgia, 91, 99
Capital, electricity, 5–6, 16–23, 54, 63, 73, 116, 160
Capital, human, 139, 141, 151, 155, 185n98
Capitalism, 16–19, 73, 167
  racial, 6, 19, 25, 79, 137, 167, 170, 175
Carter, Jimmy, 71, 97
Carver Vocational School, 145–146
Cater, Casey, 28, 39, 74
Cebul, Jason, 13
Central Atlanta Progress, 49, 153
Chakrabarty, Dipesh, 89
Chamber of Commerce (Metro Atlanta), 35–36, 49, 97, 228n85
Civilian Conservation Corps, 147
Civil Rights Act (1964), 121–122
Civil rights organizations, 25, 32–33, 44, 64, 69, 117, 121–124, 139, 146, 165, 170
Clean Air Act (1963, 1990), 96–101, 108
Clean Air Task Force, 100
Clean energy policy, 31–33, 50–55, 88–89, 109–111, 113–117, 155. *See also* climate action; Inflation Reduction Act (2022); job training; solar
Climate action. *See also* Atlanta, City of, sustainability plans; clean energy policy; renewable energies
  Climate Reality Project, 165
  Green New Deal, 14, 78, 89, 140, 141
  Kyoto Protocol (1997), 108
  Mayor's Climate Protection Agreement (2005), 51
  "Moral Call for Action on Climate Change," 165–167
  Paris Agreement (2015), 31, 60, 108
Climate change, 2, 3, 6–8, 22, 28, 31–33, 50, 52, 61, 78, 86, 88, 90, 107–111, 141, 152, 155, 160, 166–170
Climate delay, 29, 89–91, 101, 107–108, 166
Climate denial and skepticism, 3, 29, 90–91, 99, 107–108, 166
Clinton, Bill, 152
Coal, 94
  ash, 5, 80, 91, 94, 101–107, 111
  Clean Air Act (1963, 1990) enforcement, 96–101
  history in Atlanta, 92–96
  mining, 94, 98
  power plants, 90, 92, 94–96, 98, 101–107
Cobb, James C., 46
Colored Jitney Bus Association, 43

Commonwealth & Southern Corporation, 13, 41
Comprehensive Employment and Training Act (1973), 147–148
Concerned Citizens of Monroe County, 94, 105
Conservative support for solar energy, 108–111, 219–220n132. *See also* Republican Party; Tea Party
Construction Users' Anti-Inflation Roundtable, 126
Consumer advocacy, 1–3, 69–74, 80–86, 109, 116, 127, 170
  Fight the Hike campaign, 1–3, 80–86
Cooper, Melinda, 21
Counter revolution of property, 34–35
COVID-19 pandemic, 26, 55, 106
Cowan, Ruth, 21
Cowie, Jefferson, 148

Data centers, 3–4, 85, 101, 131
DeKalb Technical and Vocational School, 145. *See also* job training
Demand, electricity, 3–4, 63–69, 85. *See also* load growth
Democratic Party, 14, 91, 100, 128, 146, 148
Department of Energy, US (DOE), 59–61, 113–114
  Loan Program Office (LPO), 59–61, 76, 128
Deregulation, electricity, 15, 60–61, 75–76
Dickens, Andre, 54
Du Bois, W. E. B., 34–35, 120, 144
Duke Power, 9, 60, 93–94, 102
Durley, Gerald, 165–166

Echols, Tim, 84, 102, 109
Edison Electric Institute, 63, 107–108
Electric life, definition, 4–5, 87
Electric vehicles (EV), 3, 88, 109, 136, 156–160, 232n151
Emerald Cities Collaborative, 130, 153
Employment. *See also* job training; unionization
  clean energy, 113–117, 129–130, 137, 140–141, 155–160
  construction work, 126–130
  discrimination and, 82, 120–126, 142, 145, 146, 150, 152
  job creation, 52, 109, 128, 139, 141, 154–155, 157, 159
  job loss, 113–114, 130
  low-wage jobs, 1–2, 14, 46, 82, 151–154
  public, government as employer of last resort, 145, 146, 147–148
  safety at work, 132–133, 152–153
  unemployment, 97, 141, 144–147, 151, 153, 155, 156, 158
Energy democracy, 23–28, 30, 42, 56, 62, 86, 91–92, 111, 117, 143, 165–176
Energy efficiency, 53–54, 69, 110–111, 128, 140, 153–155, 171, 198n130, 220n137
  jobs in, 153–156
Energy insecurity, 62, 82
Energy justice, 3, 6, 24, 109, 111–112, 170
Energy Policy Act (1992, 2005), 59, 75, 76
Environmental justice, 24–25, 91–92, 100, 106, 137, 152–153, 166, 171
Environmental movement, 91–92, 96–97, 101, 106, 115–116, 129, 139, 146, 148, 170
Environmental racism, 104–106, 166, 171
Evangelical Environmental Network, 109

Fanning, Tom, 76, 87, 88–89, 97
Farley, Chandra, 54
Feminist urban political ecology, 23–28, 57, 63, 86, 90, 111–112, 163, 166–169

Ferguson, Karen, 145
Fight the Hike campaign, 1–3, 80–86
Franklin, Shirley, 51–52
Fraser, Nancy, 19

Gas (methane), 4, 54, 77, 85, 88, 92, 95, 100, 101, 111, 129, 130, 131, 171
Gender. *See also* social reproduction
  feminist urban political ecology, 23–28, 57, 63, 86, 90, 111–112, 163, 166–169
  reproductive health, 103
  women's work, 19, 21, 63–65, 92, 117, 202n32
Gentrification, 32, 45, 51, 55, 110
Georgia (state government). *See also* Georgia Public Service Commission
  Consumer's Utility Counsel, Office of, 74, 85, 172
  Department of Economic Development, 158
  Department of Industry and Trade, 48
  Environmental Protection Division (EPD), 102, 103, 106–107
  General Assembly, 38, 53, 56, 73, 74, 75, 76, 102, 106
  General Assembly, Black Caucus, 98
  Health Department, 96
  Interracial Commission, 44
  Nuclear Energy Financing Act (2009), 76
  Supreme Court, 40, 41, 71, 106–107
  Territorial Service Act (1973), 41, 73–74
Georgia Coalition of Local Governments, 54
Georgia Power, 7, 9–13, 32–33, 43–44, 49–50, 52–53, 55, 60–69, 72–74, 94–98, 101, 106, 107, 117–120, 122–125, 126, 146, 157, 167. *See also* Georgia Public Service Commission; rates; Southern Company
  business recruitment and economic development, 45–50, 135, 157–160
  Distributed Energy Resource (DER) Management System, 110, 221n143
  as Georgia Electric and Light Company, 35–37
  as Georgia Railway and Electric Company, 37, 40–41
  as Georgia Railway and Power Company, 63–64, 94
  Integrated Resource Plan (IRP), 53, 87–89, 107, 108–111, 134, 136, 171, 198n130
  worker training programs, 160–162
Georgia Power Project, 69–74, 98, 127, 170
Georgia Public Service Commission, 2, 15, 55–56, 60–62, 70–71, 73, 77–86, 168, 220n137. *See also* Georgia Power
  consumer advocacy at, 69–74, 78–86, 127–129
  labor unions and, 116, 128, 137
  nuclear and, 77–78, 102, 106–107
  as Railroad Commission, 39–40
  solar and, 107–111, 130–132
Georgia STAND-UP, 27–28, 51, 129–130, 153
Georgia Women for a New Direction, 83–84
G.I. Bill, 145, 235n38
Gilmore, Ruth Wilson, 11, 22, 105–106
Gingrich, Newt, 150
Gore, Al, 152, 165
Grady, Henry, 9
Great Depression, 12–13, 40, 144–145
Great Speckled Bird (newspaper), 69, 71, 73, 127
Green Energy Jobs Act (2007), 140
Green jobs, 52, 139–143, 146–156. *See also* job training

Green Jobs, Not Jails program, 139, 149, 153
Green Jobs Corps, 139, 233n4
Green Jobs Task Force (Atlanta), 153–154
Green New Deal, 14, 78, 89, 140, 141
Green Tea Coalition, 108–109
Grow-and-build model, 12, 29, 74, 76, 86, 96, 117, 119, 126

Hall, Bob, 73
Hall, Kwanza, 52
Hall, Stuart, 18, 117
Harambe House, 81
Harrison, Conor, 11–12, 15, 60
Hartsfield, William B., 44
Harvey, David, 20, 217n110
Health issues, 96–101, 103, 129, 169, 241n123
Henderson, Vivian, 148
Highlander Center, 94
Hirsh, Richard, 2–3, 12, 61–62
History, of electric utilities, 9–16
Home economics, 64
Housing, 7, 13, 38, 41–44, 48, 50–51, 67, 79, 82, 99, 142, 159. *See also* energy efficiency; gentrification
Huber, Matthew, 13–14
Humphrey, Hubert, 147, 148
Hunter, Tera W., 42
Hurt, Joel, 36–37
Hydropower, 13, 39–40, 92–94, 129

Income-Qualified Senior Citizens Discount program, 78. *See also* energy assistance; LIHEAP
Industrial Revolution, 17–18
Inflation Reduction Act (IRA) (2022), 4, 9, 54, 59, 113, 115, 134–136, 141, 156
Informal slating, 56
Institute for Southern Studies, 73
Insull, Samuel, 11, 181n40
Interfaith Power and Light, 165–166
International Brotherhood of Electrical Workers (IBEW), 27, 70, 81, 113–115, 119–120, 122–123, 127–129, 134, 139

Jackson, Maynard, 123, 147, 153
Jim Crow, 32, 38–45, 118–119
Job training, 46, 120, 122, 133–135, 140–163. *See also* employment; workfare
apprenticeship, 114–115, 118, 129–130, 134–135, 141, 146, 159, 162, 223n24, 224n27, 233n4
early history in Atlanta, 143–146
green jobs, 130, 139–143, 146–153
The Job Training Charade (Lafer), 141
Job Training Partnership Act (1982), 148–149
Quick Start, 156–162
Technical College System of Georgia, 145, 151, 156–162
Jones, Van, 139–140, 153, 154
Juliette, Georgia, 103–107
just transition, 114–117, 130, 137, 172

Katz, Cindi, 21
Kemp, Brian, 136, 160, 232n151
Key, James, 40, 118
King, Martin Luther, Jr., 14–15, 165
Kingston Plant, 101

Lafer, Gordon, 141, 148–149
Lake Sinclair, 96
Lennon, Myles, 19
Life-time, 21–24, 61–62, 79, 81, 90, 107–108, 160–161, 168–169, 178n16
Load growth, 3–4, 85

Longstreet, Helen D., 93
Low-Income Housing Energy Assistance Program (LIHEAP), 79, 86

MacLean, Nancy, 14
Malm, Andreas, 17–18
Manganiello, Christopher J., 92
Manpower Development and Training Program, 145–146, 156, 236n41
Marable, Manning, 10, 72–73, 104
Markets, electricity. *See* deregulation
Marston, Sallie, 21
Marx, Karl, 18
Mazzucato, Mariana, 17
McDonald, Lauren "Bubba," 81, 83, 84, 108
McKittrick, Katherine, 27, 112
Metro Atlanta Regional Transit Authority (MARTA), 45, 83, 151
Militant particularism, 106, 217n110
Mirant, 76
Mitchell, Katharyne, 21
Mitchell, Timothy, 114
Mitchell, W. E., 144–145
Moniz, Ernest, 78
Monopolization, 2, 11–12, 17, 35, 37, 60–61, 76
Mullings, Beverley, 19, 21
Municipalization, 36–37, 39–40, 73

Nagar, Richa, 27
National Ambient Air Quality Standards, 96
National Electric Light Association (NELA), 11, 37
National Labor Relations Act (NLRA), 119–120
National Nuclear Powerhouse Project Labor Agreement, 128
National Renewable Energy Lab, 53
National War Labor Board, 119
Neoliberalism, 14–15, 73, 114, 139–140, 141, 147, 155–156, 181n41, 183n66
Net-metering, 109
New Deal programs, 13–14, 67, 120, 144–149
Newman, Harvey, 27–28
New York Times, 97
Nixon, Rob, 89
Nixon administration, 126, 148
Nuclear power, 59–61, 89, 107, 128–129, 136
  nuclear construction cost recovery (NCCR), 76–78
  Plant Vogtle, 9, 59–61, 70, 72–80, 84–85, 95, 104, 116, 128–130
Nuclear Watch South, 76–77
Nye, David, 42

Obama administration, 128, 140, 154
Oglethorpe Electric Membership Corporation, 74
Olympics, Atlanta (1996), 50–51, 107

Pafford, Bobby, 71
Paris Agreement (2015), 31, 60, 108
Peck, Jamie, 14, 142
Perdue, Sonny, 158
Perry, Rick, 78
Personal Responsibility and Work Opportunity Reconciliation Act (1996), 150–151
Plant Atkinson, 95, 211n30
Plant Bowen, 74, 95, 98–101
Plant Branch, 94, 95, 96, 98
Plant Hammond, 95, 98, 100, 103, 106
Plant McDonough, 95, 98, 122–123
Plant Scherer, 70, 74, 94–96, 98, 100–101, 103–104, 105, 106, 213n55
Plant Vogtle, 9, 59–61, 70, 72–80, 84–85, 95, 104, 116, 128–130
Population demographics, 37–38, 45, 56
Populism, 39, 92
Postone, Moishe, 17, 18

Preemption, 53
Prefigurative politics, 8, 24, 167
Principles of Public Utility Rates (Bonbright), 12
Public power, 13–15, 167, 174
 municipalization, 36–37, 39–40, 73
 municipal utilities, 74, 75, 78
 Tennessee Valley Authority (TVA), 13–14, 40, 73, 101, 173–174
Public Utility Holding Company Act (1935), 13, 41
Public Utility Regulatory Policies Act (1978), 75
Pulido, Laura, 90

Qcells, 158–159
Quick Start program, 156–160

Race, 6–7, 19, 105, 139, 144, 153, 170, 235n38. *See also* job training; unionization
 Black Lives Matter movement, 55, 106, 228n85
 racial capitalism, 6, 19, 25, 79, 137, 167, 170, 175
 unions and, 115, 117–124, 129–130, 145, 224n27, 235n36
 white supremacy and racism, 12–13, 14–15, 37–45, 55–56, 93, 99, 144, 145, 223n22
Randolph, A. Philip, 121
Ransby, Barbara, 25–26
Rates (cost of electricity),
 activism against rate increases, 1–3, 39–40, 43, 68–74, 80–86, 98, 127
 calculations and setting of, 12, 29, 59–86, 107
 construction work in progress charge, 74–80
Reagan, Ronald, 148–149
Reddy Kilowatt, 63
Reed, Kasim, 52, 54, 153–154
Region (US South), 7–12, 14–15, 40–41, 45–50, 97–98, 121, 135–136, 155, 173
Renewable energies, 4, 23–24, 50–53, 107, 108, 109, 132, 140. *See also* electric vehicles; energy efficiency; solar power
Reparations, 35, 54, 144, 171
Reproductive geographies, 13–14, 63–69
Republican Party, 56, 91, 100, 106, 109, 128, 148, 219–220n132
Right-to-work laws, 113, 120–121
Rivian, 159–160
Rodriguez, Akira Drake, 41–42, 55
Roosevelt, Franklin Delano, 13–14
Rural communities, 13, 46, 48, 89, 90–94, 104–106, 118, 129, 148
Rutheiser, Charles, 33

Scherer, Bob, 74, 97, 128
Scott, Deborah, 129–130, 153
Securities and Exchange Commission, 13, 41, 108
Sellers, Christopher C., 97, 183n66
Shearer, Derek, 147, 148
Sierra Club, 31, 80, 87–88, 91, 102, 105–106, 108–109
Simmons, Fitz, 56
Situated solidarities, 27
SK Innovation, 158
Smil, Vaclav, 20
Social reproduction, 4–5, 15, 18–25, 63, 82–86, 90, 100, 116, 136, 142–143, 150–151, 156, 168, 178n16
 job training and, 136–137, 143, 158–161
 value of electric life and, 4–5, 7, 14–16, 20–25
Social wage, 6–8, 25, 120, 134, 137, 143, 151, 160, 163, 173–175
"Soft skills," 149–150

Solar power, 53–54, 107–111, 131–135, 220n136
employment, 114–115, 132–135
manufacturing, 156, 158–159
residential, 109–111, 171
Southeast Weatherization and Energy Efficiency Training (SWEET) Center, 154–155
Southern Company, 9–10, 41, 60, 48, 72–76, 84–85, 97–99, 107–108, 124–125, 128–130
meetings and reports, 16–17, 87–90, 124–125
net zero energy strategy, 52–53, 87
lobbying, 99, 100
Plant Vogtle, 9, 59–61, 70, 72–80, 84–85, 95, 104, 116, 128–130
Southern Nuclear, subsidiary, 78
Southern Environmental Law Center, 102
Southern Exposure (magazine), 126
Southern Industrial Council, 120–121
Southern Regional Council, 194n75
Southern Student Organizing Committee, 69
Southface, 153–155
Stone, Clarence, 10, 33
Strasser, Susan, 21
Streetcars, 35–45, 118–119
Stuckey, Stephanie, 52
Suburbanization, 42, 45, 48–49, 54. *See also* housing; gentrification; urbanization; urban sprawl
Suniva, 156

Tadiar, Neferti, 21–22, 62, 77
Taft–Hartley Act (1947), 120–121
Tallulah Falls Conservation Association, 93
Talmadge, Eugene, 144
Talmadge, Herman, 148
Tea Party (political party), 108–109
Technical College System of Georgia, 145, 151, 156–162. *See also* job training
Technological determinism, 24
Temporality, 16–23, 28, 89–90, 108, 112, 142, 150, 157, 165–167
life-time, 21–24, 61–62, 79, 81, 90, 107–108, 160–161, 168–169, 178n16
Tennessee Valley Authority (TVA), 13–14, 40, 73, 101
Territorial Electric Service Act (1973), 41, 73–74
Theodore, Nik, 14
Transvaal Coal Owners Association, 98
Trump administration, 31, 78, 102, 106, 108, 136, 173
Turner Foundation, 50–51
Tuskegee Institute, 143

Uneven development, racialized, 5, 25, 31–33, 54–55, 89, 99, 121. *See also* affordability, of energy
electric life and, 5–6, 16–23, 54, 63, 73, 82, 86, 89, 116
Unionization, labor, 7, 17–18, 26, 29, 43, 46, 48, 69–70, 72, 93–94, 113–137, 139, 148, 158. *See also* AFL-CIO, Georgia, Amalgamated Association of Street and Electric Railway Employees, International Brotherhood of Electrical Workers, United Auto Workers
Atlanta Federation of Trades, 119
Atlanta Labor Council, 70, 73, 127
Atlanta Plan, 123
Georgia Power, relationship to, 117–129
"Jobs for Georgians" campaign, 128
nuclear and, 116, 128–130, 136
race and 115, 117–124, 129–130, 145, 224n27, 235n36
solar and, 114–116, 134–137

United Auto Workers, 72, 136
United States (federal government). *See also* Department of Energy
  Army Corps of Engineers, 148
  Congress, 14, 99, 102, 140, 147–148, 155, 174, 213n54, 220n132, 232n151
  Department of Justice, 98, 123
  Department of Labor, 123, 141
  Electric Home and Farm Authority, 67
  Environmental Protection Agency (EPA), 96, 97, 98–99, 100, 101–104, 106
  Equal Employment Opportunity Commission (EEOC), 122–123
  Federal Communication Commission, 70
  Federal Housing Administration, 67
  Federal Trade Commission, 12
  Rural Electrification Administration, 13, 74, 173
  Supreme Court, 56, 100, 123
United Steelworkers, 115
United Way, 140
Urbanization, 10, 34–37, 48, 92, 104, 111, 169. *See also* gentrification; suburbanization; urban sprawl
Urban League, 123, 161–162, 235n36
Urban political ecology, 25
  feminist, 23–28, 57, 63, 86, 90, 111–112, 163, 166–169
Urban regime, 33, 44, 49, 54–55, 97, 113, 124, 143, 153
Urban sprawl, 45–50

Value, 6–7, 12, 16–23, 90, 116, 124, 155, 168, 175–176
  electric life and, 5–6, 16–23, 54, 63, 73, 82, 86, 89, 116
  of energy resources, 98, 110–111
  for shareholders, 60, 79, 107
Violence, 41–42, 45, 105
  Atlanta Riot, 39–41
  slow, 22, 89–90, 99, 104, 106, 111–112, 129
Voting rights and representation, 29, 55–56, 172

Washington, Booker T., 143–144
Water Infrastructure Improvements for the Nation (WIIN) Act (2016), 102
Waxman–Markey Bill, 140
WeatheRISE Atlanta program, 53
Welton, Shelley, 15
White primary, 33
Wilkie, Wendell, 13, 21
Wise, Stan, 78
Womack, Chris, 125
Women's work, 19, 21, 63–65, 92, 117, 178n16, 202n32
Woods, Clyde, 9–10
Workfare, 142, 147, 150–153
Workforce Innovation and Opportunity Act (WIOA) (1998, 2014), 142, 149–150, 151, 152, 161
Work Incentive Program, 148
World War I, 39, 63, 94, 118–119

YouthBuild, 154–155

Urban and Industrial Environments

Series editors: Robert Gottlieb, Henry R. Luce Professor of Urban and Environmental Policy, Occidental College

Bhavna Shamasunder, Associate Professor of Urban and Environmental Policy, Occidental College

Maureen Smith, *The U.S. Paper Industry and Sustainable Production: An Argument for Restructuring*

Keith Pezzoli, *Human Settlements and Planning for Ecological Sustainability: The Case of Mexico City*

Sarah Hammond Creighton, *Greening the Ivory Tower: Improving the Environmental Track Record of Universities, Colleges, and Other Institutions*

Jan Mazurek, *Making Microchips: Policy, Globalization, and Economic Restructuring in the Semiconductor Industry*

William A. Shutkin, *The Land That Could Be: Environmentalism and Democracy in the Twenty-First Century*

Richard Hofrichter, ed., *Reclaiming the Environmental Debate: The Politics of Health in a Toxic Culture*

Robert Gottlieb, *Environmentalism Unbound: Exploring New Pathways for Change*

Kenneth Geiser, *Materials Matter: Toward a Sustainable Materials Policy*

Thomas D. Beamish, *Silent Spill: The Organization of an Industrial Crisis*

Matthew Gandy, *Concrete and Clay: Reworking Nature in New York City*

David Naguib Pellow, *Garbage Wars: The Struggle for Environmental Justice in Chicago*

Julian Agyeman, Robert D. Bullard, and Bob Evans, eds., *Just Sustainabilities: Development in an Unequal World*

Barbara L. Allen, *Uneasy Alchemy: Citizens and Experts in Louisiana's Chemical Corridor Disputes*

Dara O'Rourke, *Community-Driven Regulation: Balancing Development and the Environment in Vietnam*

Brian K. Obach, *Labor and the Environmental Movement: The Quest for Common Ground*

Peggy F. Barlett and Geoffrey W. Chase, eds., *Sustainability on Campus: Stories and Strategies for Change*

Steve Lerner, *Diamond: A Struggle for Environmental Justice in Louisiana's Chemical Corridor*

Jason Corburn, *Street Science: Community Knowledge and Environmental Health Justice*

Peggy F. Barlett, ed., *Urban Place: Reconnecting with the Natural World*

David Naguib Pellow and Robert J. Brulle, eds., *Power, Justice, and the Environment: A Critical Appraisal of the Environmental Justice Movement*

Eran Ben-Joseph, *The Code of the City: Standards and the Hidden Language of Place Making*

Nancy J. Myers and Carolyn Raffensperger, eds., *Precautionary Tools for Reshaping Environmental Policy*

Kelly Sims Gallagher, *China Shifts Gears: Automakers, Oil, Pollution, and Development*

Kerry H. Whiteside, *Precautionary Politics: Principle and Practice in Confronting Environmental Risk*

Ronald Sandler and Phaedra C. Pezzullo, eds., *Environmental Justice and Environmentalism: The Social Justice Challenge to the Environmental Movement*

Julie Sze, *Noxious New York: The Racial Politics of Urban Health and Environmental Justice*

Robert D. Bullard, ed., *Growing Smarter: Achieving Livable Communities, Environmental Justice, and Regional Equity*

Ann Rappaport and Sarah Hammond Creighton, *Degrees That Matter: Climate Change and the University*

Michael Egan, *Barry Commoner and the Science of Survival: The Remaking of American Environmentalism*

David J. Hess, *Alternative Pathways in Science and Industry: Activism, Innovation, and the Environment in an Era of Globalization*

Peter F. Cannavò, *The Working Landscape: Founding, Preservation, and the Politics of Place*

Paul Stanton Kibel, ed., *Rivertown: Rethinking Urban Rivers*

Kevin P. Gallagher and Lyuba Zarsky, *The Enclave Economy: Foreign Investment and Sustainable Development in Mexico's Silicon Valley*

David N. Pellow, *Resisting Global Toxics: Transnational Movements for Environmental Justice*

Robert Gottlieb, *Reinventing Los Angeles: Nature and Community in the Global City*

David V. Carruthers, ed., *Environmental Justice in Latin America: Problems, Promise, and Practice*

Tom Angotti, *New York for Sale: Community Planning Confronts Global Real Estate*

Paloma Pavel, ed., *Breakthrough Communities: Sustainability and Justice in the Next American Metropolis*

Anastasia Loukaitou-Sideris and Renia Ehrenfeucht, *Sidewalks: Conflict and Negotiation over Public Space*

David J. Hess, *Localist Movements in a Global Economy: Sustainability, Justice, and Urban Development in the United States*

Julian Agyeman and Yelena Ogneva-Himmelberger, eds., *Environmental Justice and Sustainability in the Former Soviet Union*

Jason Corburn, *Toward the Healthy City: People, Places, and the Politics of Urban Planning*

JoAnn Carmin and Julian Agyeman, eds., *Environmental Inequalities Beyond Borders: Local Perspectives on Global Injustices*

Louise Mozingo, *Pastoral Capitalism: A History of Suburban Corporate Landscapes*

Gwen Ottinger and Benjamin Cohen, eds., *Technoscience and Environmental Justice: Expert Cultures in a Grassroots Movement*

Samantha MacBride, *Recycling Reconsidered: The Present Failure and Future Promise of Environmental Action in the United States*

Andrew Karvonen, *Politics of Urban Runoff: Nature, Technology, and the Sustainable City*

Daniel Schneider, *Hybrid Nature: Sewage Treatment and the Contradictions of the Industrial Ecosystem*

Catherine Tumber, *Small, Gritty, and Green: The Promise of America's Smaller Industrial Cities in a Low-Carbon World*

Sam Bass Warner and Andrew H. Whittemore, *American Urban Form: A Representative History*

John Pucher and Ralph Buehler, eds., *City Cycling*

Stephanie Foote and Elizabeth Mazzolini, eds., *Histories of the Dustheap: Waste, Material Cultures, Social Justice*

David J. Hess, *Good Green Jobs in a Global Economy: Making and Keeping New Industries in the United States*

Joseph F. C. DiMento and Clifford Ellis, *Changing Lanes: Visions and Histories of Urban Freeways*

Joanna Robinson, *Contested Water: The Struggle Against Water Privatization in the United States and Canada*

William B. Meyer, *The Environmental Advantages of Cities: Countering Commonsense Antiurbanism*

Rebecca L. Henn and Andrew J. Hoffman, eds., *Constructing Green: The Social Structures of Sustainability*

Peggy F. Barlett and Geoffrey W. Chase, eds., *Sustainability in Higher Education: Stories and Strategies for Transformation*

Isabelle Anguelovski, *Neighborhood as Refuge: Community Reconstruction, Place Remaking, and Environmental Justice in the City*

Kelly Sims Gallagher, *The Globalization of Clean Energy Technology: Lessons from China*

Vinit Mukhija and Anastasia Loukaitou-Sideris, eds., *The Informal American City: Beyond Taco Trucks and Day Labor*

Roxanne Warren, *Rail and the City: Shrinking Our Carbon Footprint While Reimagining Urban Space*

Marianne E. Krasny and Keith G. Tidball, *Civic Ecology: Adaptation and Transformation from the Ground Up*

Erik Swyngedouw, *Liquid Power: Contested Hydro-Modernities in Twentieth-Century Spain*

Ken Geiser, *Chemicals without Harm: Policies for a Sustainable World*

Duncan McLaren and Julian Agyeman, *Sharing Cities: A Case for Truly Smart and Sustainable Cities*

Jessica Smartt Gullion, *Fracking the Neighborhood: Reluctant Activists and Natural Gas Drilling*

Nicholas A. Phelps, *Sequel to Suburbia: Glimpses of America's Post-Suburban Future*

Shannon Elizabeth Bell, *Fighting King Coal: The Challenges to Micromobilization in Central Appalachia*

Theresa Enright, *The Making of Grand Paris: Metropolitan Urbanism in the Twenty-first Century*

Robert Gottlieb and Simon Ng, *Global Cities: Urban Environments in Los Angeles, Hong Kong, and China*

Anna Lora-Wainwright, *Resigned Activism: Living with Pollution in Rural China*

Scott L. Cummings, *Blue and Green: The Drive for Justice at America's Port*

David Bissell, *Transit Life: Cities, Commuting, and the Politics of Everyday Mobilities*

Javiera Barandiarán, *From Empire to Umpire: Science and Environmental Conflict in Neoliberal Chile*

Benjamin Pauli, *Flint Fights Back: Environmental Justice and Democracy in the Flint Water Crisis*

Karen Chapple and Anastasia Loukaitou-Sideris, *Transit-Oriented Displacement or Community Dividends? Understanding the Effects of Smarter Growth on Communities*

Henrik Ernstson and Sverker Sörlin, eds., *Grounding Urban Natures: Histories and Futures of Urban Ecologies*

Katrina Smith Korfmacher, *Bridging the Silos: Collaborating for Environment, Health, and Justice in Urban Communities*

Jill Lindsey Harrison, *From the Inside Out: The Fight for Environmental Justice within Government Agencies*

Anastasia Loukaitou-Sideris, Dana Cuff, Todd Presner, Maite Zubiaurre, and Jonathan Jae-an Crisman, *Urban Humanities: New Practices for Reimagining the City*

Govind Gopakumar, *Installing Automobility: Emerging Politics of Mobility and Streets in Indian Cities*

Amelia Thorpe, *Everyday Ownership: PARK(ing) Day and the Practice of Property*

Tridib Banerjee, *In the Images of Development: City Design in the Global South*

Ralph Buehler and John Pucher, eds., *Cycling for Sustainable Cities*

Casey J. Dawkins, *Just Housing: The Moral Foundations of American Housing Policy*

Kian Goh, *Form and Flow: The Spatial Politics of Urban Resilience and Climate Justice*

Kian Goh, Anastasia Loukaitou-Sideris, and Vinit Mukhija, eds., *Just Urban Design: The Struggle for a Public City*

Sheila R. Foster and Christian Iaione, *Co-Cities: Innovative Transitions toward Just and Self-Sustaining Communities*

Vinit Mukhija, *Remaking the American Dream: The Informal and Formal Transformation of Single-Family Housing Cities*

Cindy McCulligh, *Sewer of Progress: Corporations, Institutionalized Corruption, and the Struggle for the Santiago River*

Susan Handy, *Shifting Gears: The History and Future of American Transportation*

Manisha Anantharaman, *Communal Sustainability: Gleaning Justice in Bengaluru's Discards*

Zachary B. Lamb and Lawrence J. Vale, *The Equitably Resilient City: Struggles and Solidarities in the Face of Climate Crisis*

J. Mijin Cha, *A Just Transition for All: Workers, Communities, and a Carbon-Free Future*

Jamie Wang, *Reimagining the More-Than-Human City: Stories from Singapore*

Jason Corburn and DeVone Boggan, *Advancing Peace: Ending America's Urban Gun Violence Epidemic*

Javiera Barandiarán, *Living Minerals: Nature, Trade, and Power in the Race for Lithium*

Nikki Luke, *Electric Life: Utility Regulation and the Fight for Energy Democracy*

Publisher contact:
The MIT Press
Massachusetts Institute of Technology
77 Massachusetts Avenue, Cambridge, MA 02139
mitpress.mit.edu

EU Authorised Representative:
Easy Access System Europe, Mustamäe tee 50,
10621 Tallinn, Estonia
gpsr.requests@easproject.com

Printed by Integrated Books International,
United States of America